BEN NEWS

Teacher and fo
of Fizzics Educ

BE

Sanna Ainscough 2017

AMAZING!

How to teach science the way primary kids love

Testimonials

'This is by far the most engaging, cutting-edge and practical science book I've come across. Packed with eye-popping experiments, handy tools, great ideas and grounded in years of practical, down-to-earth educator insight from author Ben Newsome, this is a simple, fun and detailed, one-stop guide to creating an amazing STEM classroom.'

— Heather Catchpole. Editor, Careers with Code
& head of content Refraction media,
Australia's leading STEM publishers.

'An awesome new asset for teachers, written by an awesome science teacher!'

— Aaron Tait. Director of Innovation
- Education Changemakers

'I wish I had this book when I was training as a science teacher. I spent a lot of time searching for practical advice and ideas to bring the classroom to life but never found anything as useful as Be Amazing! Ben will inspire a whole new generation of science educators and their students. His expertise, passion and enthusiasm are infectious and so needed in the industry.'

— Vanessa Barratt. Science teacher and communicator

'Well done – a valuable and comprehensive guide for enhancing the teaching of science in schools and elsewhere. I'd suggest any practitioners reading the book should have a copy of the science curriculum and their lesson plan handy, and be prepared to take some notes as well as trying things out for themselves!'

– Adam Selinger. Creative Director,
Children's Discovery Museum

'Just finished your book – this is gold and a great resource! Brilliant.'

– Paul Stafford. School literacy consultant & author

'Absolutely love the document and I particularly like the common misconceptions. I use a similar approach with a True/False quizzes for my Y7 class at the start of each lesson! There are some really great ideas in this and if my Y7 class had experienced even some of these things during primary school, they would have been far better set up for science at high school.'

–George Morton. Science teacher

'Phenomenal work, I absolutely love it! It's quite apparent that the book is the outcome of over a decade of hands-on experience putting the techniques/experiments/approaches you so clearly describe into practice through Fizzics.'

–Elizabeth Fritts. Director M&A, PwC Australia

'Ben has put together a collection of experiments that make science fun and accessible and are safe, quick and easy to do. If you've been looking for a way to bring science to life in your classroom, then grab a copy of this book.'

First published in 2016 by Grammar Factory

National Library of Australia Cataloguing-in-Publication entry:

Creator: Newsome, Ben, author.
Title: Be Amazing: How to teach science the way primary kids love / Ben Newsome.
ISBN: 9780992317683 (paperback)
Subjects: Science--Study and teaching (Primary)
    Effective teaching.
    Teaching--Methodology.
    Motivation in education.

Printed in Australia by Excite Print
Cover design by Designerbility
Book production and editorial services by Grammar Factory

Disclaimer
The material in this publication is of the nature of general comment only, and does not represent professional advice. It is not intended to provide specific guidance for particular circumstances and it should not be relied on as the basis for any decision to take action or not take action on any matter which it covers. Readers should obtain professional advice where appropriate, before making any such decision. To the maximum extent permitted by law, the author and publisher disclaim all responsibility and liability to any person, arising directly or indirectly from any person taking or not taking action based on the information in this publication.

Contents

AUTHOR'S NOTE   1

GREETINGS!   3

INTRODUCTION: PRIMARY CHALLENGES
AND OPPORTUNITIES   7
    A call to action!   12
    Who am I?   18
    Over to you   21

1. UNDERSTANDING THE STUDENTS   25
    Student misconceptions: Strange ideas abound!   26
    Students' perceptions of scientists and science lessons,
    and the effect on learning   35
    Teaching science to gifted kids ... overcoming the challenges   41
    What will they remember?   44

2. PRIMING THE CLASSROOM   51
    The seed of an idea: Using stimulus material to prompt
    a discovery mindset   51
    How poor science models can still be useful for teaching science   66
    Variables and fair testing: Teaching the heart of science experiments   70
    Teacher Toolbox: Inexpensive materials that teach science at a pinch   82

3. UNLEASHING TEACHING TACTICS   93
    Scenarios that spark students' interest in science   94
    Using history as a narrative within science   105
    Create a class STEM pitchfest!   109
    Lessons and learnings when taking students on bush walks   113
    Using craft materials to teach problem-solving   120
    Building bridges using craft materials   124
    Establish a Maker space and join the movement!   132

4. LEVERAGING TECHNOLOGY   141
    Science apps for the classroom: Devices at the ready!   142
    Robotics: Teaching problem -solving in a digital environment   145
    Classrooms without walls: Break through using video conferencing!   162

Using drones to teach science   173
Some more quick ways to integrate technology into
a primary science lesson   178

5. EXPLORING SOCIAL MEDIA   189
Classroom science blogging   190
Pinterest® for education: It's a hive mind of ideas!   196
The value of short-form videos for teaching   201
Trello®: A social media solution to classroom organisation
and lesson planning   205
Teaching science through memes   212

6. ENGAGING THE COMMUNITY   221
Create a school science fair that draws attention!   222
Create a community science garden   227
Observance days and STEM learning opportunities   233
Getting creative creates community   237
Citizen science   248

YOU CAN DO THIS!   261

SUPPORT RESOURCES   267
About Fizzics Education   267
Science apps for a variety of uses   270
Video conferencing resources   283

REFERENCES   285

APPENDICES   289
Colour blindness in the classroom   289
How to improve school communication using Slack®   296

ACKNOWLEDGEMENTS   303

ABOUT THE AUTHOR   307

Author's Note

If you love enrichment and extension of science for primary students, then this is the book for you! Different countries have different syllabus requirements, different student groups have different needs. This book is not about one specific curriculum (besides, curriculums change anyway!). You know your students, you know your school and you most likely know a bunch of primary schools that already run awesome science lessons that you would do if given some support & the chance. This is your opportunity to look beyond the norm and to really set your students up for learning in later years. This is about how primary students really learn science best, it's about how to capture their imagination and how you can be that inspirational teacher they all remember. Let's have some fun!

Greetings!

'Teachers affect eternity; no one can tell where their influence stops.'

–Henry Brooks Adams

I LOVE THIS quote ... and no doubt you do too! You didn't choose teaching because there was nothing better to do. You have chosen teaching because you know the incredible impact you can have on the young lives you connect with. You have chosen teaching to *make a difference*. You have chosen to open a window to the world so that students can see how this incredible universe functions and perhaps, one day, want to teach their own children. Every day, you provide the keys for children to unlock hidden meaning in books, websites, activities, abstractions and more. It's not just about 'teaching' ... your job is to inspire!

Your work in the classroom can help bring about the real possibility of inter-generational change in science literacy. Think about the sheer number of students that you will have direct impact upon during your teaching career. What an awesome thing to think about! And if you become a head teacher, principal, consultant or similar, your curriculum decisions and the way you help others to teach can literally shape your community's attitude towards science and technology.

This book aims to look at the practical ways you can turn a classroom into a space where students are ridiculously motivated to want to learn more. A place where you can spend less time dealing with day-to-day classroom management issues and more time teaching about the world ... which is why you started teaching in the first place! Throughout this book, together we will look at ways technologies can be adapted to the classroom environment and how you can reach experts from across the globe to enhance your own students' learning. Together we will look at the very essence of the scientific method and how to teach this in ways that won't send students to sleep. Above all, this book looks at the practical ways your actions and those of your students can inspire others to approach science the same way.

Importantly, while all this guidance is valuable ... it is also meaningless without action. There are no free rides, unfortunately. But the fact that you're reading this means that you've made the conscious decision to implement more teaching ideas in your own class. Without a doubt, some of the content presented in the book will be familiar to you, while some things you might not have played with much ... which is fantastic! This means that you can mesh the new with the old and, in doing so, re-visit your tried and true lessons with a fresh frame of mind.

So, with all this in mind and a fresh cup of coffee in hand, let's delve into the ways to ignite the spark in your class and change your everyday teaching experience into something much, much greater. The best bit – it's going to be fun!

It's time to re-invigorate primary science education.

Introduction: Primary challenges and opportunities

'Somewhere, something incredible is waiting to be known.'

— Carl Sagan

THE REALITY OF full-time teaching is not quite what Sally expected. Despite all her good intentions and lesson preparations, she finds the considerable task of keeping thirty students together for seven hours a day, five days a week quite challenging.

Classroom behaviour management has become priority number one, as her rowdy bunch of Year 4 kids have a dynamic that could only be described as explosive. Sally's evenings during Term 1 and 2 have become quite predictable – long hours spent researching the topics for the next day. Lesson planning has become as much about preparation for teacher survival as it is about student learning.

She's had some awesome lesson wins (that volcano she made rocked!), but she's also had some spectacular failures (during her long, complicated talk about textiles, one student actually fell asleep!). If only she could emulate the Master Teacher she saw in her student placement! As much as she would like to balance her face-to-face time across all the subject areas, she finds herself leaning on her previous studies in the arts and literature, as this is her comfort zone, her area of strength.

Is there much science to teach? Well, she has covered the syllabus dot points as required but as the focus in the school leans so heavily towards mathematics and English, she only looks at teaching science for the minimum weekly set time… and that can only happen when the kids are listening anyway! She's fairly sure she's teaching the science correctly, but then again, there's not always someone in the room to help her. For now, she just must assume she's doing it well enough. After all, her major is in literature and drama and it's only Year 4!

Over the months, she does cover the content required in the curriculum but, with honest reflection, she never really feels in control of most of her lessons. When the end of the school year comes, however, for the most part, she's succeeded. Her first class of students has done very well under her guidance and she's offered a permanent placement at the school for Year 3, which she gladly accepts.

Does this seem familiar? While this is a fictional story, it's one I've seen repeated many of times in various guises. I've spoken with thousands of primary teachers over the years and they often recount their first teaching placements as filled with a whirlwind of ideas and potential. It can often be about survival in the classroom as you learn to corral those kids!

With classroom management being priority number one in the initial years, you become laser-focussed on keeping the students on task and if that meant playing to your strengths, well, so be it. In Sally's case, she has a major in literature and drama and so she leaned towards these subjects whenever the class needed to be kept together – completely understandable! In the first few years it can feel like such a difficult task to keep thirty students focussed on the lessons you're presenting. Now when it comes to teaching primary science, in your first years it can be quite a challenge to juggle hands-on learning materials plus get all science facts out when the kids are going nuts ... and you've got to stay on time!

It can be quite frustrating when other teachers make it look so easy, especially when they've got a few years of teaching behind them. When you watch those experienced teachers run a science lesson they seem so confident in what they're doing; they seem to be able to grab any material hanging around the room and turn it into brilliant lesson, with very little behavioural issues to boot! While you acknowledge that they've been teaching primary science for years it can still feel like it will take forever to achieve that level of mastery. Don't worry, you're not alone! It's human nature to measure ourselves against our peers. However, unfortunately, it is also human nature to give ourselves a hard time when we fall short. You need to know that this confidence you see in your fellow colleagues has taken years to craft and they too went through the same lesson failures, the same long evenings lesson planning, the same thoughts of doubt. I'd be hard pressed to meet anyone in teaching who hasn't experienced this at some point and I work with hundreds of teachers a year!

In many cases as a new teacher, you can only reflect on what went well and what didn't go well when you run a class. It takes a lot of trial and error to get it right, and your next lessons can only be as good as those extra hours spent on the internet and by getting advice from your teaching mentors as you plan your lesson sequence after school hours. Understandably you want to know more science teaching techniques, more ways of using simple materials that can be used to teach science in a way that grabs kid's attention. It's a long process and there's lots to learn, but that's why people talk about becoming a 'lifelong learner' in the first place!

It's not just about whether you feel you know enough about teaching science either. Have a chat with any teacher in a primary school about the timetable and you'll straight way hear about how little time there is to fit everything in. In Sally's case, above, the focus in her school was literacy and numeracy (for good reason too). But she also needs to fit in art, music, drama, physical education, history, geography, the school fete, parent meetings, playground duty, assemblies, lunch orders, report writing and so much more! There

are so many competing duties for a teacher to consider …
especially when many schools are being forced by necessity
to place major focus on exam preparation. Amidst the overwhelming number of responsibilities, sometimes you can feel
like you're just having to 'wing it'. Now if your background
is not in a science field, it can be hit and miss as to whether
your science lessons are working as well as they could be.

As you research ideas for science teaching you'll start to
come across confusing topic headings too. Sometimes you'll
see science referenced by itself, other times you'll see science and technology referenced together and lately there's
been the emergence of acronyms which asks you to mesh
a variety of ideas together; STEM (Science, Technology,
Engineering, Mathematics) or STEAM (Science,
Technology, Engineering, ART, Mathematics). So … does
that mean science is a singular thing to be taught or do you
always have to incorporate technology? Does technology
mean digital media only or is it meant to be a broader definition on using materials in a technological way? How does
art mesh with this? Are we meant to be teaching science on
a more holistic level or should we be diving into the detail?

When you think about it, it's quite a difficult situation to
be placed in. We all know that science is important; it im-
proves millions of lives and teaches the very basis of logical
reasoning for every child. However, it can be tough when
you've got to present science to a room full of students when

you're feeling a lack of confidence in running the content and not sure where to source all the materials.

A call to action!

'Awareness without action is meaningless.'

– Phil McGraw

Sally's concerns are ones that have been brought to me time and time again by teachers during the thousands of professional development and school incursion workshops I've delivered across Australia and globally since 2004. *How can I teach science when I don't feel confident or don't have the resources myself? Can you point me in the right direction for advice or at least to an experiment list?*

Plenty of teachers have bemoaned the lack of a plain language guide that could help them to realistically implement science in the primary classroom. Departmental policy guides and curriculum frameworks are definitely useful, but I've had many people ask for effective primary science teaching ideas – ones that are easy to implement and not only excite students but the community as a whole. I've heard the cry loud and clear – make it simple, make it more practical, make it classroom-friendly so anyone can teach it, make it less about clichéd slogans and more about what students actually want to do. In short, give us primary science teaching strategies that work!

No problem at all – you've come to right place. This book aims to help you out with this, to give you a heap of lesson ideas that have been tried and tested in classrooms many times over. This is not a formal text. It's not a literature review either. This handbook is here to help you teach the kind of science primary kids want you to teach! In this book you'll find both classic primary science concepts you'd find in standard primary science curriculums, as well as extension ideas that take your students beyond what is traditionally expected.

This is about taking students to places where they want to go!

How to teach science primary kids will love

Do you remember Professor Julius Sumner-Miller, the lecturer who first brought simple, engaging science experiments into people's homes via TV? Or perhaps Bill Nye the Science Guy? Wow, did they start something. Before their time, science in school was constrained to textbooks, some teacher demonstrations with fancy scientific apparatus, and the occasional hands-on experiment if you were lucky. Now, their idea of demonstrating science using simple materials has been taken up by primary and high school science teachers throughout Australia and around the world; in fact, you would struggle to get through a teaching degree without being exposed to this way of teaching!

What Prof Sumner-Miller and Bill Nye were actually doing is called 'science communication'. Put simply, this means communicating science clearly and effectively using simple materials so that people with little knowledge in the subject can follow and understand difficult concepts.

What has science communication got to do with primary science teaching? A tremendous amount, actually. Ignoring all the formal assessments and general duties teachers have, the bottom line is that the role of both teachers and science communicators is to ensure understanding using any technique necessary, with the ultimate goal of inspiring people to seek more knowledge than any given curriculum mandates. Both teachers and communicators must get to the heart of a science concept and build outwards from there. Beyond this, our job is to highlight how science that can be found within any subject area and that science is simply a methodical way of understanding how the world works.

This book will show you how to do just that.

This book looks at how to truly invigorate your primary science classroom, no matter what year you work with and how much time you actually get to engage in science. Rather than spending time looking into theory and policy, we'll delve deep into the ways you can quickly engage students in science in a meaningful way, where you can be sure that the take-home lessons for students amount to more than just: 'I saw a neat trick today.'

This book will take you through six key points when it comes to the primary school compass, offering practical guidance to implement as you go along:

1. Understanding the students

What's going on in the heads of the kids in your class? This chapter helps you identify and challenge the misconceptions your students are subject to when it comes to scientific ideas and to scientists themselves. On the other side of the coin, it addresses how to work, at the same time, with those children who demonstrate a more advanced understanding of the concepts than you'd expect. Overall, how can you make your lessons stick in their heads?

2. Priming the classroom

How do you set yourself up to engage students from the moment they walk in the classroom? This chapter delves into how you can use stimulus materials and science models (even poor ones!) to grab and hold the kids' attention. Importantly, it addresses how you make sure that you're creating science lessons that teach your students how scientists work and think. With discussion of how to set up the environment to encourage the kids to learn the scientific method, this chapter also provides a toolbox of materials you should keep on hand – and ideas on what to use them for.

3. Unleashing teaching tactics

Now it's time to engage with the students by going above and beyond simply addressing the curriculum. This chapter addresses different ways to inspire the kids and make the lessons come alive, from setting up scenarios and the use of historical narrative to add interest to going out into the surrounding area and using craft materials to teach problem-solving. This chapter is packed with examples of fun experiments to try.

4. Leveraging technology

It doesn't cost that much to integrate technology into the classroom and the benefits of utilising tech with students are too big to ignore. The chapter considers the value and scope of science apps, video conferencing, robotics and more. Along the way, there are case studies illustrating just what can be achieved with a little imagination.

5. Exploring social media

How you can use social media to focus students on primary science learning outcomes? It's very much the flavour of this generation, so it's certainly an area of students' lives that should be taken into consideration. This chapter runs through some of the powerful platforms you can use to your advantage. As one of their greatest influences, you can help direct your students to become better producers of useful material to share worldwide. And you'll be surprised how memes can be a tool for teaching as much as a source of humour.

6. Engaging the community

When it comes down to it, sharing knowledge can be an awesome thing for students to do, cementing their own learning in the process. If given the right structure, they'll really take it on and shine. This chapter shows how you can take teaching to a whole other level by creating events that your entire community (local and global) can get involved in!

For clarification, in this book we're focussing on science in its classic sense (experiments and the knowledge gained from them) and when I reference STEM it means that that the activity could be fit into a broader thinking of how science, technology, engineering and mathematics work together, and how they fit into the broader curriculum.

To help you with the variety of resources provided you'll find QR codes dotted throughout the book which when scanned will bring you straight to the website of interest. You can get free QR scanner apps for your smart device on iTunes® or on Google Play®.

Who am I?

'Education is not the filling of a pail, but the lighting of a fire.'

—Plutarch (often attributed to William Butler Yeats)

So who am I to be teaching you how to teach? Fair question!

I may have one the best jobs around. I run Fizzics Education, a company formed in 2004 that runs science enrichment workshops and shows in hundreds of primary and high schools. Every year, we reach 300,000 students across Australia and beyond. From my perspective, I can't think of a more varied and challenging way to use my previous high school science teaching experience! One day, we might work with thirty children at a preschool in the morning and then run a physics workshop for Years 11 and 12 in the afternoon ... knowing that the next day we might need to be at a science festival or a museum to present a large stage show to a mixed-age audience of upwards of 300 people. Some events can present quite a task when you have no idea of the audience's prior background or whether they can understand English prior to the event. Add teaching science lessons across the world via video conference into the mix and it gets quite hectic! It's a fantastic outlet after a former life as a science teacher!

Seeing the entire curriculum, from preschool through to Year 12, has given me a unique perspective on why certain concepts are in the syllabus at certain times. As two-thirds of our work is within primary schools, I've had many deep conversations with primary science specialists and experienced early educators on how they integrate science throughout their curriculum and continue to implement their thoughts on what works and what doesn't. Every day we continue to work collaboratively with schools across the country to shape how science can be best delivered to students from all walks of life. Seeing thousands upon thousands of students perform scientific investigations each year has allowed me to see how students, as learners, operate in a general sense, enabling the creation of experiments that are accessible regardless of a student's prior background. The sheer volume of students my team and I work with also forces us to think carefully about how we demonstrate particular concepts; the last thing we want to do is confuse thousands of students! Importantly, my job as a visiting science educator into schools means that I've had the opportunity to witness firsthand how teachers have created awesome science learning environments many times over.

One of the things we've discovered is that presenting science to students of all abilities and cultural backgrounds is getting easier. Science pervades the media in so many ways, from children's 'edutainment' programs through to television commercials. As science is becoming more accessible to students, science literacy is rising quickly. Years ago, I can remember presenting liquid nitrogen experiments as an enrichment visit to kids as a visiting scientist and not having a single person in the room having heard of the substance. Now, not only do students know what the substance is, but they can also suggest experiments that could be interesting for me to demonstrate. This means that presenting science to students has become a much more interactive experience, taking questions from the audience and sometimes running an impromptu experiment on the fly to investigate an issue raised. Given the opportunity and time, primary students can often understand much deeper concepts than they're given credit for and so for the past 15 years, schools have invited me into schools to reinforce syllabus learning outcomes and to extend students' knowledge.

Teaching science to students dynamically is so rewarding, and it's fantastic when I can impart this same enthusiasm to motivated educators during the teacher professional development workshops we deliver at Fizzics.

The author teaching density with simple stuff!

Where it gets interesting is that science need not be taught in isolation. There is a real interconnectedness between all the subject areas when you look deeper and it's in this framework that your lessons become much richer and more fun to deliver. For example, when you teach music it doesn't just have to be about hitting the right notes; you can also discuss the music's history, create artwork on the song's theme, write about the musician's life and home town, discuss how pitch is related to frequency, embed the music in a stage production, highlight how the rhythm in the music has patterns that can be described mathematically, and more. It's about integration of all of these ideas together!

So, the bottom line is that, armed with some tried and tested teaching techniques, you can have fun with science and be sure that the kids will enjoy it and keep wanting more!

Over to you

'The purpose of education is to replace an empty mind with an open one.'

— Malcolm Forbes

Keep in mind that as with anything you read, it will be up to you to implement the suggestions contained within this book. You may have used several of the teaching strategies to great effect and know that they work, but I urge you to challenge yourself to try as many as you can and measure the results. And by measure the results, I don't just mean a better score on a standardised test, but the real effect of positive learning in any environment that spills over into other areas.

This includes:

- Boosted student self-esteem with a clear rise in their critical thinking abilities.

- True understanding of the scientific process and the relevance of science to modern society.

- Improved behaviour and concentration levels for students across all key learning areas.

- Closer ties with the local community through your school events and parent participation.

- Increased personal satisfaction in your science lessons and new opportunities to grow professionally.

- An extended network of peers, not just within the school, but across the country and beyond.

- More control of your science curriculum and the ability to adapt to changing requirements.

- A genuine feeling that you've 'nailed it' when you finish each science unit.

There are ideas presented in this book that you can implement straight away in your own classroom, while others will take a bit of time to put together with your fellow staff. You know your students and you know what will work in your school's context. I encourage you to make notes throughout this book – grab a highlighter and pen and scribble on the pages! Add your own ideas, make checklists and timelines and stick in Post-It® notes as needed. Make it tattered, dog-eared and as abused as necessary! Reflect on what already works in your school and actively create plans with your staff to further enrich your school with both the ideas presented and others that come to mind. Be mindful about the needs of the syllabus requirements and find ways to squeeze in the extension ideas where appropriate. Don't feel that you must do everything in this book either; instead, focus on the little wins first and build from there. Keep it simple, keep it tangible and keep focussed on ever improving how you bring science into your school. Pair up with a teaching buddy and feed off each other's successes and learn from each other.

Don't expect everything to work perfectly at once; it's much better to start implementing change in your school than to hesitate through analysis paralysis (done is better than perfect!). Share these ideas and your lesson wins beyond your school so that other school communities can feed off your enthusiasm too. Action creates reaction. And, most importantly, have fun and don't let anyone tell you that it can't be done!

Time to get started? Absolutely!

1. Understanding the students

'You are always a student, never a master. You have to keep moving forward.'

— Conrad Hall

IN THIS CHAPTER, we'll look at what might be going on in the heads of the kids in your class. How can you identify and challenge the misconceptions they are subject to when it comes to scientific concepts and to scientists themselves? How do you work, at the same time, with those children who demonstrate a more advanced understanding of the concepts than you'd expect? And overall, how can you make your lessons stick in their heads?

Student misconceptions: Strange ideas abound!

'I can find in my undergraduate classes, bright students who do not know that the stars rise and set at night, or even that the Sun is a star.'

— Carl Sagan

* Just what is going on inside your students' heads?

* How much do they really know about a topic?

* Will it affect what they take away from your lesson?

So you're all set up and ready to teach students, but have you considered what misconceptions they are bringing into the room with them? It's one thing to have a science lesson ready to go, but another thing altogether to have part of your lesson plan aimed squarely at addressing the misconceptions your students have arrived with.

As many teachers and psychologists know, you understand the world through the lens of your past experiences. There is a whole bunch of literature around 'children's science', whereby, in the absence of scientific guidance from adults, kids will formulate their own interpretations of how the world works. To be honest, who could blame them? If no one is there to guide them, kids are forced to create reasons for how things work themselves. In many cases, what adults might see as basic concepts can be completely missed by kids, often producing some amusing results. In fact, these misconceptions can carry on into adulthood – I can remember being highly amused hearing a very good friend of mine recount a conversation she had a few years ago with her four-year-old daughter and her fifty-five-year-old mother-in-law while looking into the sky on a sunny morning:

My friend's daughter: 'Look how big the moon is, Mummy!'

My friend's mother-in-law: 'That's not the moon, dear.'

My friend: 'No, it's the moon.'

Her mother-in-law: 'No, it can't be, the moon only comes out at night.'

My friend's daughter: 'Yes, it is!'

My friend's mother-in-law: 'No, it isn't, dear.'

My friend: 'No, it's definitely the moon, you can see the same patterns on it like always.'

Her mother-in-law: 'Really? I never thought the moon came out during the day!'

My friend, now laughing: 'Well, yes! What did you think it was then?'

Her mother-in-law: 'Well ... I just thought it was a planet or something ...'

True story. In fact, I even checked with all the people involved when I asked if they were okay with me including this in the book! So if a 55-year-old can carry a misconception like that all these years, imagine what your students might be thinking when they enter your classroom.

While adults can get a smile from an innocent explanation that's off the mark, it does highlight the point that students will naturally formulate their own conclusions from any science lesson. This means that pointing out 'the obvious' is a critical process. Here are some key points to be aware of:

- You can't assume that the children in the room actually know what you're talking about.

- Students might have a completely different idea of a scientific process before the lesson. Remember, it's not that long ago that people thought the world was flat, that atoms didn't exist and that the universe revolved around the Earth. And there are parts of the world that still cling to these notions.

- Any prior learning could have produced a series of 'half-truths', where the kids get only part of the previous lesson and bring this into the next. This might mean that critical information from your previous science lesson is missing, bringing in the potential for the students to get lost very quickly with the new content.

The best way to pick up on these issues is to simply keep questioning the kids' understanding of a concept. Ask 'why?' as many times as possible so that students can tell you what they know. It may be worth going over old content before explaining new ideas to students, a practice tried and tested in many a classroom across the world.

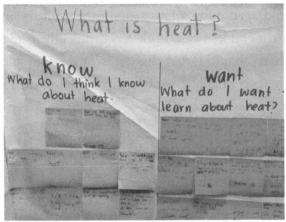

A fantastic way to find out what students want to know – Good Shepard Primary School, Rooty Hill

Why not get the students to brainstorm what they think using Post-It® notes and stick them on a large sheet of paper for a classroom discussion? This way, you can identify what students know about a topic, as well as what they'd like to know. Students can then quickly contribute to the topic, and you'll be able to tailor your teaching to the class's needs more effectively.

With all this in mind, it might be useful for you to see some of the student misconceptions that my team and I have heard while touring to present school science shows and workshops. Some of these misconceptions are well known, others very much came from left field ... but all showed us and the attending teachers that you can never assume that the students in front of you are formulating their ideas in exactly the way you expect. In no way is this a definitive list (we would need quite a few pages!); the point is simply to outline that children can think very differently to adults and your lessons should target these misconceptions and many more wherever possible. No doubt you'll have had some experience of these yourself!

Astronomy misconceptions

- All stars are the same size.

- Stars and planets appear in the same place in the sky every night.

- The moon can only be seen during the night.

- All the stars in a constellation are near each other.

- All the stars are the same distance from the Earth.

- There is a definite up and down in space.

- Planets cannot be seen without a telescope

Atmosphere misconceptions

- Rain comes from holes in clouds.

- Clouds are made of puffy things like cotton, wool or smoke.

- Gas makes things lighter.

- One degree of temperature is smaller on the Celsius scale than on the Fahrenheit scale.

Chemistry misconceptions

- Atoms can be seen with a school microscope.

- Atoms are like cells with a membrane and nucleus.

- Molecules of solids are hard; molecules of gases are soft.

- Molecules of solids are biggest; molecules of gases are smallest.

- Freezing and boiling are examples of chemical reactions.

- Physical changes are reversible while chemical changes are not.

- The original substance vanishes 'completely and forever' in a chemical reaction.

Electricity misconceptions

- An electrical circuit must be circular.

- Positively charged objects have gained protons, rather than being deficient in electrons.

- A circuit 'uses up' electricity.

- Gravitational forces are stronger than electrostatic forces.

- Batteries have electricity inside them.

Energy misconceptions

- The terms 'energy' and 'force' are interchangeable.

- An object at rest has no energy.

- Things 'use up' energy.

- There is no relationship between matter and energy.

- If energy is conserved, why are we running out of it?

Evolution misconceptions

- Dinosaurs and cavemen lived at the same time.

- A giraffe's neck got longer because it kept stretching to reach the trees.

- Humans are responsible for the extinction of the dinosaurs.

- Some human races have not evolved as much as others.

- Evolution is goal-directed, i.e. it has a purpose.

Floating and sinking misconceptions

- Objects float in water because they are lighter than water.

- Objects sink in water because they are heavier than water.

- Wood floats and metal sinks.

- All objects containing air float.

- Moving fluids contain higher pressure.

- Liquids rise in a straw because of 'suction'.

Heat and temperature misconceptions

- Heat is a substance.

- Heat and cold are different, rather than being opposite ends of a continuum.

- Boiling is the maximum temperature a substance can reach.

- Heat only travels upward.

Lithosphere misconceptions

- Rocks must be heavy.

- Earth is molten, except for its crust.

- The Earth's mantle flows quickly like water.

- Continents do not move.

- Boiling or burning radioactive material can reduce radiation.

- All radioactivity is made by humans.

Magnets and magnetism misconceptions

- All metals are attracted to a magnet.

- All magnets are made of iron.

- Larger magnets are stronger than smaller magnets.

- The magnetic and geographic poles of the Earth are located at the same place.

- Magnets are only found in toys.

- There are no natural magnets.

Properties of matter misconceptions

- Gases do not have mass.

- Air and oxygen are the same gas.

- Helium and hot air are the same gas.

- Particles of solids have no motion.

- Materials can only exhibit properties of one state of matter.

- Melting/freezing and boiling/condensation are processes only to do with water.

Knowing that the students will enter your room with all manner of ideas on how the world works, the science lessons you construct clearly need to be accurate and true to scientific understanding. But what do they think scientists do in the first place? And does this really matter?

Students' perceptions of scientists and science lessons, and the effect on learning

'A scientist worthy of a lab coat should be able to make original discoveries while wearing a clown suit, or give a lecture in a high squeaky voice from inhaling helium. It is written nowhere in the math of probability theory that one may have no fun.'

— Eliezer Yudkowsky

- Does it really matter what students think scientists do?

- Are their ideas realistic?

- Where did these thoughts come from in the first place?

35

Here's a simple experiment to run with your students. Give them a scrap piece of paper and a pen and ask them to draw and label a working scientist. No time limits, no prompts, no right or wrong answers, just simply some space on a page for your students to project their thoughts.

Look at what your class comes up with. What are the characteristics of our scientist? Is the scientist male or female? Is the scientist old or young? Are they wearing a lab coat? Are they bubbling potions or is something blowing up in the background? Does the scientist have crazy hair and a megalomaniac grin? Simply put, is there an unrealistic representation of what actually occurs in the real world?

I ran this same exercise with a class of soon-to-be science teachers a little while ago. The university students came from a variety of educational, cultural and industrial backgrounds, so you would expect all sorts of different responses. Not to be. Most students provided very similar drawings – usually a wild-haired, old, male scientist with glasses in a lab coat, mixing coloured chemicals at a bench (explosion clearly imminent if not already happening). This same exercise was put to my pre-service teaching class at Macquarie University before commencing school placements and it had the same results.

So, do your students conform to the average results we found? If so, what could that mean? To be honest, it depends on your point of view. Some people would not be concerned at all. Other people suggest that the exploitation of the 'mad scientist' image in the media has influenced people's perceptions of science; for example, within children's TV shows. You could take a cynical view and note that marketing has jumped on the science bandwagon to sell products; just look at how kids' science toys are sold or even how advertisers mark personal care products. (Personally, I'm particularly fond of the 1950's B-grade horror movie scientist.) All the aforementioned scenarios require someone, often male and wearing a lab coat, running visually appealing experiments. While I have no problem with the lab coat from a safety point of view, I do have some concerns as to why this image comes to children's minds when many scientists have nothing to do with lab coats, such as marine biologists, geologists, astronomers and so on. Why do the scientists have to be male? And, importantly, why are they 'mad'?

A study by Scherz and Oren (2005) found by questionnaire that school students' impressions of the characteristics of a scientist were consistently superficial, unreal, sometimes ignorant and often outright incorrect. This raises the question of what a child's preconception of a primary science classroom might be prior to arriving at school. Does he or she expect a mad scientist making lots of slime and exploding things in class? It seems like this is an absurd question to ask, yet speak with a few children who have newly entered Year 7 and you might find this to be exactly what they are hoping for.

Children see awesome experiments performed on TV and expect the same from their school. The students want highly engaging materials and several 'wow' experiments just to keep their attention. They want to be entertained. The problem is that if the science is just presented as a series of 'tricks' with observed results, students will naturally expect by extrapolation that the 'tricks' will only get bigger and better as they move through to high school. This occurrence is troubling and an all too familiar scenario, with students only wanting to see a 'cool trick'. In this, the student limits their understanding of the subject through not being sufficiently challenged (Shepardson et al, 2006). The reality is that, while a great primary science teacher will weave as many 'cool' experiments into their lessons as possible, the primary aim of teaching is to present science content from the syllabus in an accessible way and within the school's budget. Specifically, the hands-on experiments given to students should present an opportunity for experimentation with variables under fair test conditions ... a primary point of the scientific method.

In a perfect situation, a teacher might perform a fantastic demonstration occasionally but still recognise that students need to do their own experiments, control the variables, write scientific reports, know the reason why they did the experiments and critically observe their findings. These tasks can be unfamiliar and challenging to a student used to passive observation only, even if the experiments they have seen were awesome! If a student is running an experiment, they need to know why they are doing it, how to safely perform it and have the skills to interpret the fair-tested results, otherwise little meaningful learning is likely to occur (Hart et al, 2000).

When you think about it, all that is needed is the educator to make the science content relevant, interactive and not always a teacher demonstration. Start teaching science as early as possible and present the scientific method from the moment the first experiment in kindergarten is presented. Given a scaffold to work from, children can understand the role of controlling variables and making things fair from quite a young age. For example, if you ask a child to work out the dissolving rate of sugar in water, they might not know where to start. However, if you pose the same question with prompting suggestions on controlling heat, crystal size, water volume and spoon stirring, the child may well come up with some great results and can explain why each variable was controlled. Once the experiment has been completed, get them to write, draw or at least discuss what occurred and why. This can be done with mainstream Year 1 & 2 students, let alone the gifted child in later years. In fact, early introduction of scientific literacy activities in kindergarten children has been shown not only to increase understanding in science but also to remove the gender gap in regards to children appreciating science itself (Patrick et al, 2008). This places science teachers in a better position to get going with those excited Year 7s in later years.

Importantly, student experiment findings should be related to real world experiences, ensuring that the newly learnt concept is not left in isolation, simply to be forgotten or perceived as irrelevant. Relevance, in turn, inevitably leads to greater student engagement, thereby alleviating the need for elaborate 'tricks' to keep student attention. The ideal situation would be to present a student with a set of materials and a hypothesis, and let them decide on the best course of action to get some results. Some of the best classes we have run have been based on a question a student has posed during a lesson, where we have simply 'run with it', gathering the materials we think might be useful to experiment with and then comparing the experimental results with online research afterwards. This is active learning and a model of true scientific enquiry, whereby complex questioning based on student interests is encouraged and acted upon (Roth, 1993).

Students encouraged to explore deeper questions are generally sustained for much longer than those presented with simple rote learning scenarios or the cheap thrills of a prearranged science trick (Marbach-Ad & Sokolove, 2000). Hart et al (2000) note that even running experiments where students are not expected to understand the inherent science content has value if the experiment is designed so that students can understand how scientists establish facts.

Teaching is multi-faceted; there is no one way to run a science lesson (Buxton, 2000). Of course, if you have the resources and the time, by all means demonstrate that cool experiment you saw on TV; just make sure that this doesn't become the expected scenario in every lesson. Over time, if the primary student becomes accustomed to writing reports and evaluating findings, they will find the transition to the high school science laboratory an extension of their primary studies, rather than a foreign experience.

In the meantime, it's important to realise that, while many kids will enter the classroom full of misconceptions, there will be times where your students will be light years ahead of their peers, both in understanding the content and their ability to draw predictions and conclusions that are well beyond the level of the rest of the class. So how do you deal with this in the primary science classroom?

Teaching science to gifted kids ... overcoming the challenges

'What makes a child gifted and talented may not always be good grades in school, but a different way of looking at the world and learning.'

— Chuck Grassley

• How do you balance competing needs across your classroom?

- How do you pitch lessons at multiple levels?

As most readers will be aware, working with gifted children offers its own challenges when it comes to science. My approach to any child, gifted or not, is to work with what they can do rather than concentrate on their particular age. Having supplementary material on hand is the best insurance to keep active minds busy, as well as being flexible enough to alter the lesson if the workshop tasks have been aimed too low or high. Simply, from my perspective, I've found that working with gifted children just means that you practise good teaching methods rather than concentrating on the tag 'gifted'.

Still, even being used to working with different groups of gifted kids, I can be surprised by what some of these children have as prior knowledge. I remember running a human body workshop with some gifted Year 2 children and mentioning the various cell types found in the bloodstream. A child promptly noted that I should be calling red blood cells erythrocytes and that there are several types of white blood cells which use phagocytosis, the process of foreign cell engulfment, when fighting bacteria! Investigation into the child's background revealed that the child's father was a haematologist and the child had been looking through his dad's medical school texts. Needless to say, the lesson objectives for this child were shifted to the content you might find in this year's high school biology exam!

Solving the human body puzzle in Seven Hills
North Public School

The issue is that gifted students often don't have access to opportunity classes and are generally placed within the mainstream student population. It is all well and good to change a lesson's direction if you can, but this is often at the peril of losing the focus of the other students, who are also capable of understanding the concept but need more time (a precious commodity). Solely catering to the gifted child will mean that the others miss out, which is not really equitable. So, what to do? Try to anticipate where a lesson might go when working with gifted children. Have some materials set aside that the gifted kids can work on while the rest of the class works on the core lesson concepts. In other words, be prepared to run 'multiple lessons' within the same lesson … a tough task, I know! But if you spend the time preparing a great unit of lessons that cater for all abilities from the outset, you'll have greater engagement across the class, regardless of ability. If you don't end up using that extra lesson material you prepared, you can always use it in the next lesson. The best bit is that classroom behaviour will improve across the board ... and your job gets easier!

What will they remember?

'I can't tell you how many people say they were turned off science because of a science teacher that completely sucked out all the inspiration and enthusiasm they had for the course.'

— Neil deGrasse Tyson

- So you've taught the lesson; what did they get out of it?

- How do you balance 'WOW' experiments with the little things?

Nothing puts your lesson to the test more than having students try to recall just what the content was. You can tell how hard it is if you ask them to send you a drawing, especially if you ask for it to be sent a week after the fact, or, better still, ask them questions about what they learned a month later. The trap of running a science lesson is that you can get caught up in the 'whiz bang' demonstrations, the fancy gear and the theatrics involved, and lose sight of the learning that's taken place.

I've seen many school science shows and workshops in my time (who would have thought?). Generally, they can be split into several categories, all with parallels in the primary classroom:

- The show pony – fancy dress, fancy props, high energy, in an all-singing, all-dancing extravaganza.

- The travelling teacher – student-centred and curriculum-oriented, less concerned with shiny, pull-up banners and pretty props and more concerned with spreading knowledge.

- Some combination of the above.

- The 'I should be doing something else' burned-out teacher … we've all seen this one!

Whether it's running a science show on stage or teaching kindergarten about hot and cold, what matters is knowing the audience you're presenting to. You have to ask the question … what are you trying to achieve?

I've seen all manner of science lessons pitched perfectly at the right group of people and others where the wrong thing has come out on stage or in the classroom. You cannot be everything to everyone; it's the context that matters most. The kicker is that if you truly want to teach for a living, you need to walk the fine line between being an entertainer, a teacher, a storyteller and, occasionally, a crowd control officer!

Sugar, skittles and water on a plastic plate is a
beautiful experiment – no need to enhance it!

No matter what age group you're teaching, please consider
the science that you're representing; someone had to develop
it, someone has applied it, someone respects it and, finally,
your audience has to understand it. My suggestion is to be less
concerned about doing the 'flashy experiments', at least not
when you first start out, and be more concerned about ensuring the students you have can follow your explanations and
that the science is correct. The aim is to have your class truly
learning something new, something that they will bring back
home to share. Aim the content at the right age level and
speak in their language and context. If you achieve this, then
you've done your job, which is representing scientists and their
work, coherently communicating the theory and hopefully
inspiring the next generation of thinkers. Most importantly,
what did your kids take away?

ACTION POINTS

- Take the time to learn about your students' misconceptions. That goes for every lesson, as you'd be surprised what they're thinking. It really comes down to the age-old teaching tactics of good questioning and class brainstorming prior to the main body of the lesson.

- Find out what your students think scientists look like and what they think they do. It is highly important that they know that the clichéd 'mad scientist' image is simply fiction and just a bit of fun. Not all scientists wear lab coats. Not all scientists are male. Not all scientists are old. Scientists work in nearly every industry and are there to solve problems and create solutions for the world to use.

- Have a secondary lesson plan as a backup in case of gifted students. Even if you don't use it, at least you'll have an extra lesson prepared for a later date. At the very least, you'll now have a deeper understanding of the topic itself.

- Think carefully about how the science is presented in class. If it's all 'WOW, BANG, POP!' and no substance, you'll be selling an image that just doesn't exist. On the other hand, if it's just dreary, tired experiments, you'll turn the students off as well as yourself. Aim for a balance between great teacher demonstrations and decent hands-on activities. Let them feed off your enthusiasm and even the simplest of activities can be a winner.

Need help?

At Fizzics Education we're working with Education Changemakers to invigorate science in primary schools via professional development courses. Drop us a line to find out more!

Notes

_____

_____

_____

_____

_____

_____

_____

_____

_____

_____

_____

2. Priming the classroom

'Tell me and I forget. Teach me and I remember. Involve me and I learn.'

— Benjamin Franklin

Now we've looked at the kids, let's have a look at the classroom. How do you set yourself up to engage students from the moment they walk in the room? How do you make sure that you're creating science lessons that teach your students how scientists work and think? It's your classroom after all; let's make it awesome!

The seed of an idea: Using stimulus material to prompt a discovery mindset

'If a child can't learn the way we teach, maybe we should teach the way they learn.' Ignacio'

— Nacho' Estrada

- Is there anything in your classroom that promotes critical thinking and reflection?

- Is your classroom a sterile, unchanging place or is it inviting and vibrant?

- Do you have anything to ignite students' curiosity?

Ever wondered what is going through the kids' minds when they first enter your classroom? Apart from the usual chatter about what they did on the weekend, what happened on TV last night and the standard hustle to get their bags packed away, kids are also both consciously and subconsciously looking around your classroom to see what you've got in store for them. This begs the question ... what do you do to grab their attention when they first walk in the door? Let's be honest, as a teacher, the first things you're often dealing with are getting kids to hand in their homework, settling them down, marking the role and maybe sorting out a playground incident that happened before the bell rang. This means that you have at least five to ten minutes' worth of 'downtime' in your classroom every day that is effectively taken up by the nature of classroom management. Now when you think about it, that's effectively between twenty-five and fifty minutes of face-to-face teaching time per week that students miss purely because it takes time to settle the class and start your day. So, how can you claw back that lost time and still deal with the commotion that is the start of a school day?

I was dropping off one of my kids at the local childcare a while ago and the first thing I saw in the entry corridor was a setup of flowers placed into a test tube rack. This was quite neat. At the time, Kmart® was selling these for around $7 and they made a nice decoration to brighten up the sign-in desk where we dropped off the kids. It got me thinking, though. What if this same arrangement could have a longitudinal science experiment set up to demonstrate that plants have vascular tissue that transports water up the stem to the leaves and petals? Both parents and children go past this desk on the way in; wouldn't it be great if they could watch the petal colours change over the coming days if you added some food colour? This is a standard botany experiment using white flowers run in both primary and high school science lessons the world over, and it can easily be adapted for early learners too.

Flowers at my child's day care centre

So what has this got to with the beginning of a school day? Well, imagine the increased interest your students would have in the world around them if you constantly changed their learning environment by adding stimulus materials to the classroom. When the students sit down, they're going to be looking around the classroom anyway. Why not have something set up where the students are actively encouraged to quickly check out what you've placed on a desk before they sit down? The students could then spend five minutes writing down their observations regarding what they think the lesson might be about in a daily record book for later discussion (this could later be placed in your classroom blog too – more on that later). Meanwhile, you could be handling the possible playground incident or collecting homework or covering any number of the myriad of things a teacher is expected to do prior to running a lesson!

Of course, any suggestion such as this will take some initial planning as part of your lesson sequence and a genuine commitment to make it happen, but the payoff could be well worth the effort. If students become accustomed to you requiring them to critically look at and engage with the stimulus material you prepare each day, over time you will reduce the amount of lost lesson time at the start of each day. More importantly, you should begin to see the creation of a discovery mindset, where what you're actually asking the students to do is observe the world around them rather than just blunder into class mindlessly. By the time you're done marking the role, most of your students will have at least begun the process of 'switching on', which means that you can concentrate on getting the remainder of the class on task far more efficiently ... all of which saves your time in the long run!

Constantly changing stimulus material in your class is simply another way of practising teaching at its best. You don't have to be predictable either; some weeks you can change the setup on a daily basis for some quick experiments and other times you could set up something that lasts a week or two (very handy for when you're busy teaching other things). Also, you don't necessarily have to be the one who always sets up the experiment or stimulus materials – why not assign groups of students in a rotating roster where they create the experiments for others to observe? The kids will see that you actively care about their learning, as it will be obvious you want them thinking from the moment they walk in the door. Also, if you plan your stimulus material to be directly related to the science lesson you're about to run anyway, you'll be getting a head start on the lesson and gaining their attention without even trying. Now what is there not to love about that?

Some quick ideas for stimulus material in your classroom

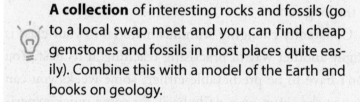 **A collection** of interesting rocks and fossils (go to a local swap meet and you can find cheap gemstones and fossils in most places quite easily). Combine this with a model of the Earth and books on geology.

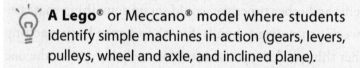 **A Lego®** or Meccano® model where students identify simple machines in action (gears, levers, pulleys, wheel and axle, and inclined plane).

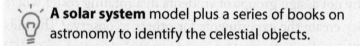 **A solar system** model plus a series of books on astronomy to identify the celestial objects.

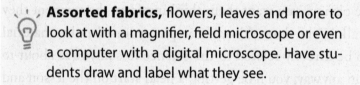 **Assorted fabrics,** flowers, leaves and more to look at with a magnifier, field microscope or even a computer with a digital microscope. Have students draw and label what they see.

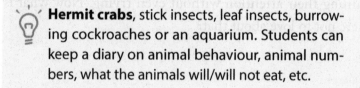 **Hermit crabs**, stick insects, leaf insects, burrowing cockroaches or an aquarium. Students can keep a diary on animal behaviour, animal numbers, what the animals will/will not eat, etc.

A biological model (human body, flower, insect. While some of these can be expensive, you can still get some useful ones at the toy store or online). Again, support material such as books or perhaps a computer opened to a support website would be handy here. Students can draw and label what they see.

 Natural selection demonstration – have assorted jelly beans on the desk with a sign: 'Only take one jelly bean to eat, this is important!' Which flavour of jellybean was most popular? Which flavour 'survived predation from students'?

 Density column showing different coloured liquids floating on top of each other (e.g. various oils, water, glucose syrup, etc.). You could even float different objects throughout the column, depending on their density too.

 A lemon battery, with two different metals pushed into a lemon and connected to a multimeter to show the electricity flowing (you can get a multimeter quite cheaply from a hardware store).

A Cartesian diver made out of a floating pen lid weighted with Blu-Tack® in a plastic bottle filled to the top with water. Add a sign that says: 'Squeeze me!'. Students can explain why the pen lid sinks when the bottle is squeezed.

 A lava lamp, where students explain the motion of the oil as it heats up.

 Anything that you're about to teach about that day!

10 long-term experiments students could set up in your classroom

You'll see that I've included variables to test in these – this is critical and I go into more detail on this later.

Description	Materials needed	How long could you run this?
Grow some sugar crystal rock candy!	**Teacher uses:** ■ Kettle ■ Water ■ Sugar **Students use:** Cups, ■ Pegs, ■ Wooden skewers, ■ Cooled sugar solution **Variables to test:** ■ Sugar concentration ■ salt vs sugar	1 to 2 weeks Full experiment description: http://tiny.cc/ fizzics-rock-candy

Description	Materials needed	How long could you run this?
Make crystal snowflakes	**Teacher uses:** ■ Kettle ■ Borax from the laundry section of your supermarket **Students use:** ■ Cooled Borax solution ■ Popsicle sticks ■ Pipe cleaners ■ Scissors **Variables to test:** ■ Borax solution concentration	3 days to 1 week Full experiment description: http://tiny.cc/ fizzics-snowflakes

Description	Materials needed	How long could you run this?
Make an ant farm	You can buy ant farm kits online these days quite easily or you could set one up in a thin fish tank or two sealable jars so you can see the ants burrowing away. Choose a soil with a bit of texture so the tunnels don't collapse (a sandy loam would be ideal). **Variables to test:** ■ Soil texture ■ Ant species (obvious safety point here – don't choose ones that bite!)	1 week to 1 month Full experiment description: http://tiny.cc/ fizzics-ant-farm

Description	Materials needed	How long could you run this?
Bean growth experiment	**Students use:** ■ Re-sealable plastic kitchen bags ■ Tissue paper ■ Water and beans placed near a window (tape onto a window for great results) **Variables to test:** ■ Light vs darkness ■ Water vs salt water	2 to 3 weeks Full experiment description: http://tiny.cc/ fizzics-beans
Make your own biosphere	**Students use:** ■ A 2-litre PET bottle ■ Scissors ■ Sticky tape ■ Soil, pebbles ■ Charcoal fragments ■ Small plants **Variables to test:** ■ Types of plants ■ Types of soil ■ Amount of charcoal ■ Light vs dark ■ Cold areas vs hot areas	1 month or more Full experiment description: http://tiny.cc/ fizzics-biosphere

Description	Materials needed	How long could you run this?
Brine shrimp *Artemia* spp.	Also known as Sea-Monkeys, brine shrimp can be found online or in a pet shop in a dormant state. Place in salted water and watch them hatch and grow with magnifiers! **Variables to test:** ■ Try observing their behavioural responses to light vs dark. Brine shrimp are highly sensitive to water quality – be nice to them!	48 hours to 2 weeks Full experiment description: http://tiny.cc/fizzics-brine-shrimp
Make coloured petals and visible vascular bundles	**Students use:** White flowers + stems, celery stalks, containers with different food colourings + water **Variables to test:** ■ Plant species	2 days to 1 week, dependant on plant species and temperature/humidity

Description	Materials needed	How long could you run this?
Make a naked egg	**Students use:** ■ An egg ■ Vinegar ■ A cup **Variables to test:** ■ Strength of vinegar	One week or more. Full experiment description http://tiny.cc/ fizzics-naked-egg
Rust a nail	**Students use:** ■ A nail ■ A cup ■ Salt water ■ Vinegar ■ Oil ■ Coke **Variables to test:** ■ Types of nails ■ Size of nails ■ Different concentrations of salt ■ Salt water vs vinegar & similar. Try comparing nails vs steel wool!	1 week to 1 month Full experiment description http://tiny.cc/ fizzics-rusty-nail

Description	Materials needed	How long could you run this?
Time-lapse photo diary of your school garden ... 1 photo per week at the same time throughout the school year	**Students use:** ■ A camera ■ Pen ■ Paper **Variables to observe:** ■ Flowering ■ New leaf growth ■ Fauna ■ Shadow direction ■ New seedlings	1 year Full experiment description: http://tiny.cc/ fizzics-garden

You could also set up a class weather station and measure wind speeds, humidity, temperature, wind direction and rainfall, and plot them over a month or more. I've also seen primary classes use lux meters from electronic stores to plot sunlight levels as well as students observing carefully-sealed bread in plastic containers to watch mould growing over several weeks (safety point – tape these containers carefully and don't open them; simply throw them out after the experiment!).

Getting children to critique stimulus material helps greatly when it comes to increasing understanding. Critiquing science models can have the same effect. As you pull your science lesson together, you'll start to notice scientific inaccuracies in all manner of things ... even in the science resources you find in your own storeroom! Don't despair, though; it's quite possible to teach science with objects in your classroom even if they're completely terrible in design.

How poor science models can still be useful for teaching science

'The scientific theory I like best is that the rings of Saturn are composed entirely of lost airline luggage.'

— Mark Russell

- Are there scientific models in your classroom that could be critiqued?

- Do your students have the skills to do this?

This might be counter-intuitive but, sometimes, presenting poorly-designed scientific models to students can be extremely helpful when it comes to allowing students to critically analyse a given topic. My favourite version of this is to find the worst possible model of the solar system that I can, then get the students to tear right into it!

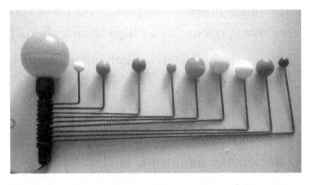

Solar system model: terrible ... but useful!

Of course, as you'd expect, at Fizzics, we make sure we bring the most accurate and exciting science models to show the students. However, there are times when it's best to choose exactly the opposite. The contrast between good models and bad can be so great that students can't help but want to get involved pulling apart the bad science. With the solar system pictured above, it's great to have students brainstorm all the things wrong about the model ... and believe me, there are a *lot* of issues:

- The size ratio of the sun compared to the planets is completely out.

- The distances between each planet and to the sun are totally wrong.

- Where are the moons? For that matter, what about the asteroid belt, the Kuiper belt or the Oort cloud?

- The sun is missing sunspots, solar flares and is even the wrong colour (it's a yellow dwarf, after all).

- Mercury is completely smooth instead of pockmarked with craters.

- Venus should be white-ish yellow, reflecting the presence of a carbon dioxide atmosphere laden with sulphur.

- Earth has no continents, tundra, weather systems or polar caps.

- Mars is missing its northern polar ice cap.

- Jupiter has no gaseous banding or rotating storms shown (and nor do any of the other gas giants).

- Saturn is missing its icy rings, and that's the same for Jupiter, Uranus and Neptune.

- If we are only showing planets and our sun, why is Pluto there if it's classified as a Dwarf Planet?

- The orbit paths are completely circular instead of elliptical ... and the list goes on!

Now, of course, you could call this being picky (it is, in reality, a $30 model), but that's the point – if students can critically analyse how bad a scientific model is, it means that they really do know their content and can demonstrate scientific literacy. In the case of this model, you could easily spend forty-five minutes or more going to town on how bad it is and, I must say, in several cases, I've had to stop the students so we don't run out of time for other things! The aftermath of an intense forty-five minutes of student interrogation is usually two whiteboards completely filled with highly informative facts and figures about what should be included in an ideal model of the solar system ... and the teacher could not be happier. In a short space of time, the students are taught to question what is presented to them, form their own opinions, research the facts and collectively contribute to a better model that they would prefer ... thereby emulating thought processes used in the development of new ideas, as well as undertaking teamwork around a common goal.

What's the moral to this story? Even if you can't get the exact scientific model to demonstrate a concept, the very flaws you find can be the greatest source of inspired learning. In fact, you can use a lesson like this as a scaffold for students to begin to question everything around them.

By using stimulus materials and models, you've got the students' minds ticking over about the science topics they're going to study. Now let's look at how you construct that topic in your classroom. How can you ensure that they learn about the scientific method itself? It comes down to how you get students to view the experiments in the first place.

Variables and fair testing: Teaching the heart of science experiments

'Science is a way of thinking much more than it is a body of knowledge.'

— Carl Sagan

- Are you teaching science or just teaching a series of tricks?

- Can your students identify what factors affect your experiment?

- Do your students know when the results of an experiment are fair?

While this is not the most 'fun' area of science to cover, it is one of the most important aspects of science teaching that needs to be addressed if you're going to ensure that your class understands how the scientific method actually works. Whether you're teaching science to primary kids, secondary students, pre-schoolers or undergraduate students, your ability to convey the importance and the skill of fairly testing variables is critical for learners taking their first steps into the scientific world. On our visits to schools, we've often found that people request a simple overview of how teachers can teach scientific methodology in their classrooms, regardless of the subject at hand.

In plain language, the scientific method simply boils down to a researcher's ability to identify a question that they want to ask, where they can use controls to modify just one aspect of a given situation so that measurements can be taken and a valid conclusion can be drawn. This typically involves a step-by-step process:

1. Identify the things going on in a given situation (for example, temperature, humidity, light levels, plant height and so on). We'll call these 'conditions' from now on. Get students to brainstorm what they already know about the situation and the conditions that you've identified to create a pool of the class's prior knowledge.

2. Determine if any of the conditions above can be changed at all. From now on, we'll call changing 'varying'.

3. If one of the conditions can be varied, can you accurately measure this condition? One example of a condition that could be varied would be temperature.

4. Determine whether the rest of the conditions can be controlled (to continue the previous example, you might put everything in a greenhouse). If so, proceed further. If not, do you think it matters? To put it another way: Would your results be open to someone saying that your experiment was done poorly due to

there being something affecting your experiment that you should have taken into account?

5. Knowing that you can accurately modify a condition and measure it while you control everything else, is there anything about the situation that you would like to find out? What I mean by this is: Can you create an experiment where you change a measurable condition to see if it has any effect on another measurable condition of the experiment (for example, does plant height get affected by temperature?)?

6. Make a prediction, based on your prior knowledge, regarding how you think varying this condition will affect the other condition (perhaps you think that plant height increases as you increase the temperature).

7. Run the experiment against a control case, whereby one set of conditions never changes at all (for instance, the plant stays at the same temperature for two weeks), while at the same time running another set of conditions where one of the variables is changed over the same given timeframe and you measure the effects (temperature is increased on a regular basis over the two weeks and plant height measured).

8. Record the results accurately each time you do the experiment. Preferably, replicate your study lots of times to get rid of any mistakes you might have made.

9. Draw a conclusion based on your results.

10. Write up the experiment so that someone else can reproduce what you have done using your instructions to see if they also prwoduce the same result.

What is the situation?

Can you change one thing to test?

Can you keep everything else the same?

Can you measure the changes you make fairly?

Can you measure the effect of that change?

Make a prediction on what will happen

Run the experiment and include a control

Accurately report all of your results

Form a conclusion from your results and research

Can someone repeat this experiment accurately?

Yes?
Share it!

No?
Fix the experiment and
go to step 1

Now, after explaining the above to your students and letting them get used to the idea, you can start to introduce scientific language. Change the word 'condition' to 'variable' for a start, and highlight that the word variable comes from the word 'vary', meaning to change. Once the students come to grips with the fact that you have variables in an experiment, it is time to introduce the types of variables without confusing them. Personally, I've found it easiest to ask the following:

- What did we vary each time in the experiment?

- Did our results reflect us measuring the variable we changed during the experiment?

- What other things did we control in our experiment?

In the above example, the students will quickly identify that you were changing the temperature each time and hoping to see a change in the plant height, with everything else remaining constant in the greenhouse. Now it's time to put a list up on the board of the above questions, except this time give them their scientific names:

- The independent variable is the variable that doesn't get affected by other variables. This is the one we vary each time during an experiment.

- The dependent variable is the one we measure for a potential effect from changing the independent variable (i.e. it *depends* on what you do to the thing you change).

- The controlled variables are the other variables we keep constant during an experiment and don't expect a response from when changing the independent variable.

Simple sentence for kids to remember

As the scientist changes the independent variable, the effect on the dependent variable is measured. Control the rest!

Some slight repetition and now the students can see where you're coming from. You're almost there! Ask the students if they are happy with how the experiment went. They might say yes/no/maybe/not sure, but you can then rephrase and ask directly, 'Was the experiment fair or not?' Kids have an inherent idea of what is fair; you just have to watch them argue in the playground over taking turns! Have them evaluate the experiment and work out whether the way you ran the experiment allowed you to fairly test the question you raised, such as: 'Does temperature affect plant growth?' To help them, you can ask them a variety of questions:

- Did we accurately measure the temperature? Was there anything that could lead to an error here?

- Did we accurately measure the plant height? Was there anything that could lead to an error here too?

- Did we accurately control the other variables? Was there anything that could lead to an error here?

- Was there anything else when we replicated the experiment that gave a result that could be considered unfair?

This is a good time to discuss why you had the students run the experiment several times. Why? To get an average reading across the experiments to reduce experimental error. At this point, you've more or less nailed it – by having students follow the sequence of identifying and controlling variables and accurately measuring the result, all while posing questions as to whether the experiment is fair or not, you'll go a long way towards setting up the mindset needed to run any experiment in any discipline. What you're actually teaching is experimental design, a critical thinking process that should never be skipped prior to running an experiment in the real world. A great scenario to pose to students about the value of good experimental design is this: Imagine if you spent three years and thousands of dollars on a particular study only to find out at the end that your experiment was flawed from the beginning. You'd be more than upset!

Finally, you need to mention that the point of writing up the experiment is so that someone else can repeat it the way you ran it and check if they get the same results. In scientific speak, you're creating a falsifiable and repeatable experiment. In other words, are you just making stuff up or can someone test the validity of your claims? This reminds me of a very famous quote:

> *'No amount of experimentation can ever prove me right; a single experiment can prove me wrong.'*
>
> — Albert Einstein

With this in mind, below is a simple scaffold of the basic titles you see in a typical experiment report at a Year 5/6 level. Of course, different schools have different opinions of these title names, but the essence is more or less there. As students enter high school, potentially, they'll be asked to put in an abstract as a requirement and have their introduction include more researched detail (a must in university, by the way)

Title:
What is the experiment about?

Abstract (optional):
Summarise the whole report as quickly as possible.

Introduction:
What is the background information the reader should know about (often skipped by students)?

Method:
What materials did you use in the experiment and how did you use these materials?

Results:
What data did you collect during the experiment?

Discussion:
What is the significance of the data results? Do the results confirm the background information research you did or have you found something new? Explain, explain, explain!

Conclusion:
What is the bottom line of the experiment?

References:
What textbooks, websites and more did your students get their background information from?

In practice, the above structure is often flexibly delivered in primary schools and is tailored to the abilities and learning needs of the kids anyway (we can be rigid about this later in school life!). Sometimes primary schools ask that students specifically state their hypothesis in a separate heading towards the start of student report, which is fine as this teaches kids to think carefully about what they are studying and design the experiment to address the experimental question raised. Other primary schools also get the students to fill out an extra section stating the exact variables that they are working with so students can get practice in identifying them. Unfortunately, I've found a number of students who have learned to skip the introduction, which is not that handy, as this is the student's chance to do a little research (which is an important skill) and pass this onto the reader. When you think about it, in primary school, it could be as simple as a quick paragraph or two on what they've found out in library time.

Regardless of how you go about teaching students to write their reports, it's more important that they understand why they are doing it, the reason being ...

to communicate effectively so others can repeat the experiment exactly to test your claims!

Now is the time to reinforce the concept by posing an entirely new experiment and guiding them through identifying the variable types, undertaking fair testing and, finally, evaluating what they did. If you repeat this exercise enough times, the thought processes of a researcher will be ingrained in your students and the whole process of how scientists work will be less of a mystery. They'll then be able to identify variables quickly in any given experiment and pose a testable question that produces a valid result that can be repeated by someone else.

At the same time, it's definitely worth chatting with students about how they could apply the scientific process to their own lives:

- What is the best ratio of ingredients to use in a chocolate cake?

- Which shampoo gives your hair most strength?

- Which lawn fertiliser is the best to use?

- Which octane level produces the best fuel economy in a car?

For the vast majority of situations you're looking at, there's often a way of scientifically checking what is actually going on. Essentially, the scientific method is saying: 'If I have a situation with elements X, Y, Z, what would happen to Z if I varied X and controlled Y? And how can I fairly test this to record a valid answer that is reproducible and testable by someone else?'

If you follow all of this and begin to apply it in your classroom, the kids' minds can only grow as a result. It was the very advent of the scientific method that really accelerated civilisation to produce our current way of life. Instilling in our students an understanding of how scientists work is a very important task for all teachers, regardless of whether the students are academically gifted or not. Unfortunately, not doing so risks creating a generation with a lack of understanding of scientific processes. This can quickly bring about public mistrust and even animosity towards the very people working hard to use verifiable evidence to improve our lives. Besides which, wouldn't it be great to have the skills to work out why things happen all the time? Sounds like a plan to me!

Of course, while planning to run science experiments properly, you also need to consider what resources you have on hand. These resources don't have to break the bank; they just need to be effective.

Teacher Toolbox: Inexpensive materials that teach science at a pinch

'Rockets are cool. There's no getting around that.'

— Elon Musk

- How much science can you teach using simple materials?

- Is there anything in your craft cupboard you could put to good use?

Having a 'quick grab box' of science materials can make the world of difference to any number of teaching scenarios. Imagine you need a quick reward for good student behaviour or that your planned lesson runs much shorter than you expected and now you've got thirty students to keep occupied for the next thirty minutes until bell time. Storing away some non-perishable items in a plastic box does wonders for when you need to produce an engaging lesson at a pinch. This idea has long been advocated by experienced casual science teachers, as they know full well they don't always arrive at a school with everything set up for them, and they now need to control Year 9 for a double period! It'll help you in primary science teaching too.

To create an on-demand 'science go-to box'; I recommend getting a sturdy, 5-litre, plastic box with a lid so you can easily carry it up flights of stairs. To make this box as versatile as possible, you should source science materials that can be used in multiple scenarios and have enough of the materials to run at least one experiment where all the students get to have a go. Knowing this, it would be worth considering adding the following to your 'science go-to box':

Balloons	Stapled tea bags	Blue-Tac®
Wooden kebab	Water bombs	Tissues
sticks	Bicarbonate	Rice
Plastic straws	soda	Flat, shallow
Nylon string	A small bottle of	plates
Paper	vinegar	Food
Scissors	Salt	colouring
Matches	Funnel	Small plastic
Candles	Ping pong ball	Cola bottle &
8mm metal nuts	Pipe cleaners	lid
Small bottle of	Needle	Plastic Zip-
detergent	Plastic cup	Loc® bags
		Magnet
		Sticky tape
		Photographic
		film canister
		(you can still
		get these!)

At this point, you might be wondering what on Earth you might do with the above science materials.

Well, you could use a combination of the above materials to run every one of the science experiments listed next:

Just some of the experiments you could do with simple materials from your 'science go-to box'

Use the search box on Fizzics website or simply visit http://tiny.cc/fizzics-experiments

- Fill a balloon with carbon dioxide
- Make a milk rainbow (just grab some full cream milk from the staff room fridge)
- Film canister rocket
- Make a volcano
- Balloon and nut hummer
- Make a straw flute
- Sticky static balloons
- Bernoulli balloon blow apart experiment
- Make different coloured petals (you just need to collect some white flowers

- Balloon survives the flame
- Ping pong on string
- Soap powered boat
- Tea bag rocket
- Make a lung model
- Air takes up space demonstration
- Upside-down water cup experiment
- Make a simple compass
- Skewer a balloon
- Impossible puff
- Geometric bubble films
- Make a mobius strip
- Demonstrate friction using rice

If you're unfamiliar with any of the experiments noted above, take a look at our website www.fizzicseducation.com.au. Here, you'll find pictures and instructions that you can print out and keep in the box with your materials.

Without a doubt, there will be some readers who look at the list of materials above and add even more science experiments to the list! The materials don't have to be ground-breaking. Sometimes even the simplest things can explain important concepts.

You'll find that most of the materials are hanging around your school anyway, and the rest can be picked up at the shops quite easily. The best bit is that nearly all the materials are re-usable, which means that the act of creating your 'science go-to box' effectively means you've saved time in the long run when it comes to future lesson plans. Given just how many science experiments you can create from these materials, you'll have several days of science lessons on hand!

AN EXPERIMENT IN SIMPLICITY

Teaching about plant structure by raiding the fridge

This one seems like a simple task to run with your students, but you might be surprised by the results. Next time you're planning on running a basic plant anatomy class, why not pop down to the shops and pick up some grocery items from the fruit and vegetable aisle to test them with? You'd expect people to quickly work out if they're looking at a seed, inflorescence, stem, berry, fruit, tuber or root, but it can actually be a little tricky ... especially if the terminology used at home and in general conversation is filled with misnomers.

A quick introduction to the structure of plants can be done using a pile of common fruits and vegetables on the presenting desk. Get the students to vote on each item as to what it is. No definition needs to be given for each classification at this stage, as the point is to survey the class to see what they think. The following table gives the class spread of results from an after school science club I once ran for Year 5 and 6 students:

Tomato	60% fruit, 20% berry, 20% vegetable, 5% seed and 5% unsure (Answer = berry)
Potato	85% root, 15% tuber (Answer = tuber)
Corn kernel	80% fruit, 10% berry, 10% seed (Answer = seed)
Broccoli	80% inflorescence, 15% stem, 5% seed (Answer = inflorescence)
Banana	95% fruit, 5% seed (Answer = berry)
Carrot	80% root, 15% tuber, 5% unsure (Answer = root)
Cabbage	85% leaf, 15% flower (Answer = leaf)
Celery	95% stem, 5% leaf (Answer = stem)

As you can see from the above, the class did pretty well, but there was definitely some confusion, especially with the banana, the corn kernel and the tomato. Of course, as we progressed through the items, the kids started to pick up on the similarities between each of them and were swayed by other more vocal opinions in the group. It was interesting that not one of the items got 100 per cent of the class to make the right call.

Of course, the difficulty in the task above lies in how you define each plant part (this varies from source to source). The main problem was in working out the fuzzy notion of what constitutes a 'fruit' as, botanically, there are several types of fruit.

Since a fruit is an organ that contains and protects seeds, you could pop tomatoes in either the 'fruit' or the 'berry' category in this activity, and you could argue that both answers are right ... until you find out that fruit is the broad category and that a berry is a much more defined structure in botany which covers the tomato much more closely. This makes the definition of a banana as a berry quite strange to students and causes quite some contention in the group. Interestingly, the students also struggled to define a potato as a tuber as opposed to a root ... this was due to some of them not knowing the word 'tuber' and others trying to second-guess the quiz!

The issue is that if kids are not exposed to the correct biological language from outside the classroom, they're going to bring a variety of preconceptions into your class that you need to address. I know I grew up thinking that a tomato was a vegetable as it went into salad; you might find that many of your kids will think the same thing. In fact, I'd challenge you to run the same test with some adults! So to help out, here are some simple definitions that might help your next class on plant anatomy. Of course, it would be wise to consult an up-to-date biological description as, every now and then, the definitions change.

If you really want to get them thinking, why not introduce the words 'drupe', 'nut', 'petiole' and more ... but perhaps baby steps first, especially for primary students!

ACTION POINTS

- Make your classroom stimulating! Change stimulus material often and get students into a routine where they learn to critically observe experiments over long periods of time. Soon, it'll become routine and you'll be cultivating minds to actively engage with their surroundings.

- Don't be afraid to point out the flaws in your teaching materials. Sometimes, it's those very weaknesses that can be your biggest strength. Get students to critique scientific models for inaccuracies. When coupled with solid research, this can be a powerful teaching tool.

- When creating your lesson plans, make sure that you include sections on identifying variables and hypothesis testing, as well as discussions on whether the experiment was fair. This way, you'll be teaching the heart of the scientific method and students will become accustomed to thinking this way in no time. Otherwise, your science lessons will effectively just be a series of tricks. Teach students to write up the experiment thoroughly so someone can repeat the experiment to test their claims.

- Raid your cupboard for craft materials and create a grab box of science materials. Include print outs from our website so that you have pictures and instructions for what to do (www.fizzicseducation. com.au).

Notes

_____

_____

_____

_____

_____

_____

_____

_____

_____

_____

_____

_____

3. Unleashing teaching tactics

'A fool is a man who never tried an experiment in his life.'

— Erasmus Darwin

So, YOU'VE TAKEN the time to understand your students, you've reviewed and planned your science lessons to address the scientific method, you've got stimulus material in your classroom to critique, and you've got a bunch of inexpensive materials ready to teach science at a pinch. This is fantastic, as you're really in the perfect position to get students to understand how their world works. Now it's time to engage with them by going above and beyond simply addressing the curriculum – to really make your teaching come alive!

Scenarios that spark students' interest in science

'The greatest way to live with honour in this world is to be what we pretend to be.'

— Socrates

- Can you tell stories to grab the imagination, no matter the age of the learner?

- Why not get a little creative?

Scenarios always grab the attention of students, firing up their imagination. And creating a scenario that challenges students to solve a mystery or puzzle is always great fun. When this is combined with a concrete timeline to get the study done and a requirement for the students to present their findings in character, you've got yourself a lesson that kids will remember for years! It's all about forming a narrative that is convincing for kids that puts them in the shoes of a researcher as they go about finding the clues in a series of hands-on experiments. There are so many subjects to consider: Gold mining in the 1850s, medieval siege weapons, climate change simulations, agriculture used by indigenous cultures, creating a roller coaster park and much more. I've even seen teachers put together lessons around a pretend zombie apocalypse!

What all of these lessons have in common is that they place the learner into the middle of a role play, whereby you challenge them to ask questions and explore their own creativity as they work within the realms of science to solve a given situation.

One of the classic ways that science teachers have been using a scenario to teach science is via a forensics murder mystery. What student wouldn't want to try to solve a crime using techniques they've seen on CSI, Bones, NCIS or any other primetime television program? A forensics workshop is a natural fit for student groups, enabling them to rotate around stations, and it easily fits into most school science curriculum units, so much so that we've been running our own forensics school incursion visits for years. Still, as engaging and fun as it is, think about whether a forensics unit always has to be about a gruesome murder. As you might imagine, you have to carefully choose whether this narrative is right for your students (this could bring up bad memories for some kids, unfortunately). Even if your students and your school are completely fine with you running this lab, at some point, you may feel that you want to change it up a little ... even if it's just to keep yourself interested!

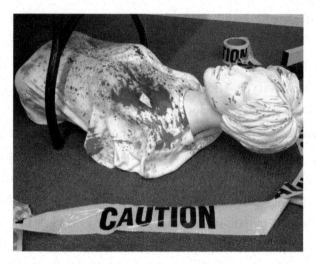

This scene would certainly grab attention! Still, be sure to know the background of your kids prior to a setup like this.

With this in mind, here are some extra sleuthing ideas that you can run in your classroom. Try to make each piece of 'evidence' logically work with the available materials. When presenting tasks for the students to do, they'll appreciate the effort you've gone to in re-creating a pretend puzzle to solve! I've placed a few ideas under each forensics scenario to get you started; each evidence type could simply be set up in a series of stations that the kids rotate around. The more you surround your students in the immersive theme, the better the class!

1. **Shipwrecks and salvage** ... which vessel have the marine archaeologists discovered?

a) **Handwriting analysis:** Do the records from the retrieved 'ship's log' match the captain's handwriting found in letters sent home? Get another staff member that kids are less familiar with to write the messages so that they seem authentic (make the writing as elegant and elongated as possible!).

b) **Under the microscope:** Does the recovered 'sail' fabric match known historic samples of the original sail manufacturers? You can get different material samples from home or a fabric shop.

c) **Under the magnifier:** Have samples of wood from previously recovered 'wreckages' to compare with that of the vessel the kids are studying. Aim to get different softwood and hardwood pieces to work with ... this could be harder for the students to work out but it's also an opportunity to mention that a 'null result' is still a result.

d) **Cannonball bores:** Does the size of the recovered cannonball match known sizes of the cannonballs used by the ship? You can use school shotputs and some measuring tape to produce a realistic and fun way for kids to apply some mathematics to the lesson.

e) **Mapping analysis:** Does the shipwreck's location match historical records of where it was known to be sailing? This could be interesting to put into the puzzle, as you could state the 'known average speed' the ship sailed at, the time the boat left the port, the location of the port and the location of the shipwreck on a map. Match to the ship's log for dates and location.

f) **Cargo contents:** The ship was carrying a consignment of black ink from China (also known historically as India Ink), but does the cargo ink match the ink known to be transported? This sounds fancy, but you could use this as a lead into chromatography, where you place ink dots on filter paper and then dip the edge of the paper in water ... this causes water to pass through the ink dot and draw out the colours, which can then be matched to the evidence.

2. **Who stole the parrot?** Our prize-winning parrot has gone missing ... trace the clues to find the culprit!

a) **Blood analysis:** The criminal cut themselves on the birdcage ... can you match the blood type of the criminal to one of the suspect samples? Mix red food colouring and chocolate topping for each of the blood samples, but add a little full cream milk to the evidence and the sample of the suspect you want to frame. The students then react the blood samples with vinegar (posing as an antiserum) to see if the clumping pattern matches.

b) **Ransom note analysis:** What clues are there on an extortion note found in the birdcage that can help track the criminal down? Here you could apply the chromatography and handwriting ideas used in the shipwreck scenario plus perhaps some latent images … when you write on a notepad, you leave indents on the pages underneath (place some clues that link the suspect to the extortion note!).

c) **Footprint analysis:** Inside and outside the room, there are muddy footprints … can you match the prints to any of the suspects' shoes? Footprints are easy to make on flat surfaces but you can also get students to pour Plaster of Paris into muddy footprints to get a cast to check out too. Just be sure to wear a face mask and gloves when you prepare the Plaster of Paris and follow all safety directions on the packet.

d) **Feather detective:** Match the feathers found in the suspects' homes. A simple trip to the petshop can provide you with a bunch of feathers to use under a microscope or magnifier!

e) **Security footage:** Is there anything in the images taken by the cameras that might reveal the identity of the thief? You could make these photos grainy and poorly-lit with a couple of manipulations using Adobe Photoshop® or similar.

3. **Cryptic codes** ... someone has hacked your computer but left code traces as to who they are.

a) This is one for the students who love codes. You give the kids cipher wheels, cryptic keys and more to have them decode a number of made-up computer scripts where you have inserted messages. This could be time-consuming initially but, once done, you'll have a great series of mathematical puzzles for students to sink their teeth into.

b) **An alternative idea** along the same thinking as the above would be: **Wow, you've just heard that there is a hidden message in a song! Can you decode it?**

4. **Which bug is this?** The museum needs your help to bring order to their collection chaos.

a) In this case, the actual forensics is simply applying a dichotomous key. You might remember these as the flow charts you used in a high school, which had a cascading series of questions with yes or no answers (e.g. is the beetle green or brown?). The students sort out the museum chaos and learn a bit of biology in the process. You could apply the same lesson plan with rocks too!

As you can see, there are quite a number of ways that you can completely change the topic and still run a class based on solving a mystery. When it comes down to it, you can easily put together a variety of science experiments to fit a teaching scenario; it just depends on your creativity, time and access to resources.

Teaching science through a space mission scenario

When I was on my Churchill Fellowship a couple of years ago, I was privileged to visit the Challenger Learning Center in Rochester, New York. The team here took immersive student scenarios to a whole other level by having students participate in a two-hour experience, placing half the class in a pretend spacecraft on a mission to Mars and the other half in a simulated Mission Control back on Earth.

101

Connected by video conference, the students had a number of STEM challenges thrown at them against the backdrop of a countdown to landing, all the while surrounded by an environment that looked and felt authentic, even down to students wearing costumes.

In the command centre, the kids worked in small teams on requirements including spaceship navigation, medical, remote robot guidance, data communication, life support systems, energy systems, probes, and communications between the ship and ground team. In the spaceship, the students had other science experiments and spaceship guidance activities to perform. The pace could only be described as frenetic as, in real time, the students completed activities that were critical for the success of the mission; teamwork here was the key. So many aspects of STEM were covered: Simple circuits, geology, magnetism, energy, human biology, mapping, robotics, acid/base chemistry, code breaking, the solar system … I could go on for a while. The best bit is that the Challenger team had woven pop culture throughout the presentation, combining jokes and music to not only liven up the presentation but also highlight the need for levity in stressful situations. During the mission, the kids swapped so they each got their turn in the two spaces – this was done very well and did not seem to be jarring to the experience. Simply fantastic!

In the Challenger Learning Center, Rochester NY

Can you adapt this to your classroom? Why not! Even if it's watered down to suit your budget, the students will still get a kick out of it.

The team runs another student experience at the Bathysphere Underwater Biological Laboratory (BUBL), where, again, the students are placed in a station-based exercise. This time, they are trying to solve the question as to who has been polluting Lake Ontario, a fun environmental take on the forensics theme. This is connected via video conference too.

Of course, as a classroom teacher, it is unlikely that you will have the time and resources to produce this level of detail every time. However, scenarios can still be run effectively on a smaller scale. And maybe you could get students to create a scenario for another class to solve? If the students can make their own scenario and use it to teach others, it will truly demonstrate that they have mastered the knowledge of the subject. Another option is to plan out the lesson sequence with your colleagues so that you share ideas and tasks between you. You could treat it like a showcase, where each class chooses one aspect of the scenario to focus upon, and decks out their classroom with decorations and experimental apparatus to show the rest of the school. Imagine the buzz in your school if you pulled this off! The rewards are great for those who dare.

As you can see, setting the scene for kids to explore is, in many ways, a no-brainer for increasing classroom engagement. In fact, the very act of creating a scenario means that you have to define the time in which the scenario is set. This means that you can weave history into the equation to produce a richer narrative and bring more context to the students, bringing them to an understanding of why the inventions and discoveries occurred in a certain time and place.

Using history as a narrative within science

'Philosophy of science without history of science is empty; history of science without philosophy of science is blind.'

— Imre Lakatos

- What's the point of science if the students don't know why discoveries were being made in the first place?

- How can you place the students in the story?

I was once running a home school science workshop on electrical circuits when, during the session, I got into a discussion with one of the kids' parents around what their children could be reading beyond standard classroom texts. As we explored this further, it became clear that the real issue was that, while the textbooks they were using were scientifically correct and beautifully laid out, the language itself was just too formal to be engaging. This type of writing is, of course, understandable, considering that the aim of the text is to inform and guide with facts, but the feedback from the children was that the classroom texts just didn't capture their imagination; a couple of kids even found them boring, despite the authors' and illustrators' best intentions. The bottom line – there was just no compelling narrative and context to capture the kids' attention.

The best example of the opposite end of the spectrum is an old favourite of mine: Bill Bryson's *A Short History of Nearly Everything*. Nominated for the Samuel Johnson prize, this book takes the reader through the initial period of star formation after the Big Bang to discussions about the Earth's formation, atmosphere, microbiology, evolution and more. Importantly, the writing is story-based; you can read about feuds between North American palaeontologists, the struggle to reach our deepest ocean floors and why we even bothered, interviews with astronomers on finding supernovae from backyard observatories, the deft diplomacy Edmund Halley used to get Isaac Newton to publish *Principia* after a falling out with Robert Hooke, the race to find the impact crater from the meteorite that contributed to the demise of dinosaurs, even the opinions people held on Edwin Hubble, plus the weird fixations exhibited by Carolus Linnaeus in naming plant structures. It's not just a science book; it's filled with political intrigue, big personalities and interesting examples on just how difficult it can be for scientists to change long-held beliefs, even among their peers.

Bill Bryson's book is actually a version of a narrative known as 'Big History'. What's Big History all about? It's an attempt by noted academics and well-known authors to pull together information across all the sciences into a continuous story that begins at the Big Bang and works its way to now, with each chapter building upon knowledge raised previously. In threading these chapters together, the reader is taken on a journey to a deeper understanding of how the sciences are interlinked and how natural events and human discoveries throughout history have impacted on today's society. Because the information is presented in a timeline rather than discrete chapters, the reader can follow how the presented ideas work together and make their own linkages in the process. More importantly, as historical events are filled with drama and personalities, you can find yourself learning why a certain discovery came about, where and when the thought arose, and the impact on subsequent events. In many ways, it reminds me of the role-playing strategy computer games I grew up with, where you guided people through history so they could survive whatever the millennia threw at them (remember Civilization® and Populus®?). Of course, the text within Bill Bryson's book is going to be beyond many primary students however it's worth a read as a teacher so that you can see how he threads science and history together and then consider how some of this could be done at the right learning stage too.

There is a global movement to address STEM using Big History, and there are now plenty of web-based resources on hand that could help support you should this be of interest. It would be worth checking out the Big History Project, which has a strong following and even stronger backing. Developed by Professor David Christian of Macquarie University as a college subject, it has turned into a non-profit organisation designed for teachers and students around the world, with support from Bill Gates. Filled with animations and well-structured lessons for learners, the aim is to mesh STEM with History at a curriculum level. As a teacher, it might be worth you having a look at the developed units of work and support resources, and considering whether this might be something that could be offered as an extra subject at your school.

One of the best ways to address waning interest in textbooks is to give students a reason to look further in the first place. Let's be honest, no one really likes reading a bunch of facts that have to be remembered to complete a worksheet or exam. Why not combine elements of Big History with an engaging learning scenario, as mentioned previously? Students generally want to be inspired; they just need to be convinced that gaining knowledge is worth the effort. As educators, it's worth searching beyond classroom texts to find information sources that have been written deliberately to provide meaning and context, not just pretty pictures or rules to remember.

Both creating scenarios in the classroom and giving students some historical context are great ways to invigorate your teaching. But why not use your surrounding environment to teach as well? Beyond your school, there are many parks and gardens throughout the suburbs and towns. There's bound to be an area nearby that you can use to immerse the students in the natural environment. If you're stuck in a concrete jungle, it's worth considering taking your students out of the city and into the forest, even if it's just for half a day.

Create a class STEM pitchfest!

'Innovation distinguishes between a leader and a follower.'

— Steve Jobs

- How can you instil a sense of entrepreneurship into your STEM classes?

- How can you teach students to think about creatively solving problems in a way the world could use?

The business pitching TV show Shark Tank® and its competitor Dragons' Den® have certainly captured the public's imagination in the past year or so. Sitting in your lounge room, you can imagine what it would be like to walk in to a room with venture capitalists and spend time pitching your business concept to experienced entrepreneurs, who will not only critique your concept but perhaps also offer funding and business support in exchange for equity in your business concept. It is highly engaging television and has seen winners go on to grow their businesses and profile substantially as a result of going on one of these shows.

Why not take this idea into your primary science classroom and create a STEM pitchfest? Not only will it be fun to see what your students come up with, but the students will also learn multiple skills:

- Sticking to design briefs and allotted times.
- Using creativity to apply science concepts.
- Presentation and persuasion skills.

Like any lesson, this will take some resource planning and time for students to be taught some background science so that they have a scaffold of knowledge to build from. Before you send them on their creative journey, it would be only fair to lay down some ground rules about how the competition will be judged. Here are some tips:

- Share the marking rubric first. This seems like a no-brainer but if you miss this step you're bound to be accused of being too subjective. Lean the points towards the STEM concepts raised by each design, but still have room for creativity and the ability to persuade the audience on the need for the idea.

- Ensure that they know this is for fun! You're not having them compete to win real funding, but if you introduce this lesson well they'll take it very seriously all the same, even if it's just to win a chocolate bar!

- Consider having a balanced judging process that includes a student panel too. If you do so, consider holding a lottery between each student presentation to rotate the judges. Give them all a marking rubric to work from as well. For classroom management, you might need to be careful about breaking up cliques here.

- Another option for judging could simply be via a student poll. Poll Everywhere® (polleverywhere.com) is a great service for running polls and works across multiple devices. The students can then see the answers get collated in real time. Great for those BYOD (bring your own device) classrooms!

- The class doesn't necessarily have to be modelled on Shark Tank® or Dragons' Den® (those shows, after all, are about business first and foremost). You could instead take the model of the ABC's New Inventors, which is directly about inventors trying to bring their concept to market.

- Have the students create a working model of their idea, complete with a budget for the materials and how long they have to create. Your students could blog about the design process and list the scientific theories their ideas encompass.

The reason for this lesson is to get students to start thinking about the applications for the scientific discoveries and engineering oddities that are made every day. Some ideas may not yet have use but they are still important to investigate regardless. This could bring about the debate on the economics of research; there has been a shift in modern times towards research with a potential fiscal outcome, but is this a good thing or a bad thing? Students often don't see the point of certain areas of science and now is the time to show them!

You could also use this science lesson as an opportunity to talk about pseudo-science and outright quackery. In the early 20th century, there were snake oil salesman and people swindling money out of investors for their perpetual motion machines. Sadly, a well-presented pitch can still dazzle people and the facts can be left behind in the rush to get in first on a breakthrough. Even to this day, this occurs. Only recently, a couple of years ago, the major biotech company Theranos® burst onto the scene with a promise to revolutionise blood tests; this year, they got into hot water around their medical blood testing devices' efficacy and subsequently lost billions of dollars in value. At the time of this book's publication, Theranos® is scrambling to sort out the science behind its product. The lesson for students is that you need thorough independent auditing and review before you go to market.

Furthermore, the idea of science and public perceptions of a particular field of study could lead onto a variety of deep conversations where you could talk about climate change, evolution, vaccinations, plate tectonics, the universe and other contentious issues that create friction among the general public. Just be careful what topic you choose and the age group of the students involved. You don't want to leave them with a half-formed misconception!

As you can see, a simple lesson around pitching can spark a strong and robust debate with your students. As long as you carefully manage your students' emotive responses, you can be assured that they will begin to think more broadly about the important place of STEM in our everyday lives. Perhaps you'll inspire a student to create the next *big* thing in science and technology!

Lessons and learnings when taking students on bush walks

'The jarring change going from an urban environment to an extremely remote natural environment is extremely inspiring. It's constantly stimulating, it's like a slap in the face.'

— Carter Burwell

- How often do your students experience the natural world?

- Do your students know more about faraway land-scapes than their own backyard?

The great outdoors offers so many learning opportunities for students. I can still remember my Grade 6 teacher taking us down to the local creek in Brisbane as part of an environmental studies unit. No more than ten minutes' walk from the school,

113

all of us were very excited to be able to walk along the bush paths in the hope of finding the animals and plants that we had been learning about back in class. During our visit, we spent time learning about the impact of suburban runoff on weed growth, why the vegetation type changed as we entered the gully, where to look for possum drays and much more. We also collected leaf, water and soil samples to analyse back in class, and our teacher took photos of where the samples were taken for future reference. In no time at all, our student groups had built a body of material to use in both our project reports and the upcoming debate on how to remediate urban bushland settings.

Unfortunately for some kids, going on a bush walk with their teacher represents one of the few times in their lives that they are allowed to head 'into the wild'. This means that taking them outside your classroom is your chance to inspire them to understand their local environment and to take notice of the world around them. Of course, you need to do your due diligence in respect to student safety, but the pros really do outweigh the cons.

Walking in Lachlan Swamp, Centennial Parklands NSW

Here are some teaching opportunities to discover when you head up that trail:

- **Look for the little things.** You'd be amazed at the amount of biodiversity to be found in leaf litter. Just have the kids wear gloves, and use tweezers and sorting trays. For those looking at relative abundance, you could do a comparison between leaf litter found underneath different tree species or between completely different vegetation types. For those wanting a bit more rigour, you could bring a quadrat to create discrete sample sizes (if you don't have access to one, you can knock one up in about ten minutes using wood strips from the local hardware store).

- **Speaking of little things,** bring along some ice cube trays, spoons, dip nets and larger sorting trays, and you could spend quite some time sorting through a variety of aquatic organisms found among the vegetation alongside water. If you have an identification book with you, it can very much help with those tricky IDs – we use Williams' *Australian Freshwater Life – Invertebrates of Inland Waters* (MacMillan Education, 1980). We have an awesome time doing this activity when running holiday science programs at Centennial Parklands. Try scooping among the reeds and riparian vegetation for better results (good for a comparison actually ... where is the highest number of living things found and why?).

- **Take readings** of relative humidity, temperature and more if you're ascending a mountain. You'd be amazed how the data points change on bush walks, and it's these physical parameters that affect plant and animal distributions. You can get measuring materials quite easily from most science supply stores or even via a trip to your local electronics shop.

Using a hygrometer in a rainforest

- **Go spotlighting!** This is great for an overnight camping trip with students or even a special science evening. Keep to a well-formed track for safety, and bring along a torch to look in the trees for arboreal mammals. Possums are a classic to find in the Australian bush, but you might find even more reclusive creatures. This activity can be really engaging for kids; you'll just need to highlight the importance of being quiet or most animals will be hidden well before you get to them. If there is a nearby stream in the right area, you might be able to look for platypus too.

- **While spotlighting,** you could also listen for frogs. Bring along a smart device to record the sounds and then work out what you heard back at the classroom. This activity is ideal if you happen to have a naturalist with you, but if you don't, there are apps that can help you too (check out the Australian Museum's frog field guide, for instance).

- **Keeping on the same track** as listening for frogs, you could also do the same for bird life. Bring along a pair of binoculars and a local bird identification book to help – we use *The Field Guide to the Birds of Australia* by Knight and Pizzey (Harper Collins, 2005).

- **Incorporate some mathematics!** For example, you could measure the height of trees using a clinometer. Simply squeeze the trigger and you get an angle of your eye to the top of the canopy. With a bit of trigonometry, you can gain a relatively accurate idea of the height of your tree stand. There are apps for this too!

Measuring heights of trees with a clinometer

Extra safety tips:

- Let people know your start time and expected return time.

- Bring extra food and water, sunscreen, insect repellent, plus children's medications.

- Bring a first aid kit, a whistle and several ways of communicating with people (lesson opportunity here!).

- Be aware of weather conditions and bring a map.

- Stick to the trails.

- Don't push beyond your students' limits.

As you can see, there is no shortage of teaching opportunities when taking students out into the environment. Start with the end in mind – think about what they are going to learn and what data you can use when you get back to class. If you carefully balance the learning opportunities with the simple pleasure of being outdoors, your students will appreciate the natural environment more and your future lessons will have much more context.

If you can't get your students out into the wild, you can still get your environmental fix by having a park ranger, zoo keeper or similar incursion provider visit your school for a lesson in ecology and biodiversity. This can save on the paperwork needed when taking students out of school grounds and saves money on buses too!

So, we've looked at running scenarios, involving history and entering the environment as ways to engage the kids' interest, but what if you're more concerned with the technology and engineering aspects of teaching? Again, there's a simple answer: Spend time with your class using craft materials to make simple machines or solve a specific challenge in a short timeframe.

Using craft materials to teach problem-solving

'The walls between art and engineering exist only in our minds.'

— Theo Jansen

- Students learn best with their hands, especially with the right incentive; can you give them a competition to win?

- Can you start to look classroom craft materials in a different light?

Teaching science in your classroom doesn't necessarily only have to be about following the textbook. If you want to truly engage kids in the sciences, you have to tap into their creative side. Let them go about solving a scientific problem their own way and see where their discoveries take them.

The recurring issue in the classroom is not having access to all the materials you want; you can get discouraged when you hear about the awesome things being done by schools with bigger budgets. The good news is that a lack of fancy resources does not have to be a limiting factor. You'll find that as long as you engage the students with a question worth solving, they are just as happy working with simple materials as with the expensive stuff. This thinking is the bedrock of the Maker Movement, where students tinker with all sorts of materials to produce a creative solution to a problem. Lately, there has been an inclination to relabel STEM, taking into account the mix of disciplines, and call it STEAM instead, acknowledging the artistic processes inherent to creative design challenges.

Beyond learning to work with materials on a design challenge, a major outcome for STEAM projects in the classroom is that kids can learn through failure. While it can feel good for students to have continual success in their experiments, this doesn't teach them about the real world, where you often have to deal with challenges when not all the variables are known. In fact, teaching students to participate in STEAM experiments where the outcome is not guaranteed is incredibly important, as learning to operate in these conditions is critical when pushing the boundaries of knowledge. It is an important component of the discovery and entrepreneurial mindset.

So, with that in mind, what are the kinds of STEAM challenge you can run in your classroom on a limited budget? The following is just a short selection of classic primary science favourites:

Water bomb or egg drop challenge

This is a classic engineering challenge that students absolutely love. The basic premise is that, given a limited time-frame and a limited materials budget, students have to make their egg or water bomb survive a free fall. The goal is to test your student's creativity in a scenario that engages the students' interest; in this case, we suggest having the students pretend that they're landing an astronaut and sensitive equipment on Mars. You can also liken the challenge to the engineering issues faced when it comes to car safety, where seat belts, crumple zones, the roll cage, airbags and more are constantly tested by road safety experts to ensure passenger protection. The main safety concerns are making sure that kids are not allergic to eggs (if so, use water bombs instead), and ensuring that you've got a safe place to drop the contraptions from a reasonable height.

You'll need to set up some rules that everyone must adhere to around the following:

- The types of materials the students can use. Generally, any sort of classroom materials hanging around will suffice, but the following might be a good starting list: Balloons, string, plastic cups, rubber bands, tissue paper, wooden kebab sticks, drinking straws, paper clips, sticky tape, cotton wool balls and so on.

- The amount of materials they are allowed to use. Be strict about the material use, as you are getting them to model what happens when you have a budget (very important in any job!). You could even introduce bonus points for using fewer materials; not only would they be 'saving money and resources', but you could also discuss how reducing payload weight means that the rocket needed to lift their craft would need less fuel.

- What happens if they break materials (for example, snapped rubber bands, popped balloons, cut drinking straws). Up to you here, but I've seen classes respond really well when they know that they need to be careful with their materials. For example, if students choose to chop up the straws, they stay chopped. This means they have to plan what they are doing and they have to adapt to the consequences of material failure, both useful skills to have!

- A hard deadline for launch. Not only is this good for lesson planning, it models the real world, where the space agencies have very specific launch windows due to the positioning of celestial objects. I tend to only give the egg or the water bomb to the students at the last moment. Not only does that stop mess from the items being dropped at the wrong time, but it also requires them to design a hatch or cargo bay door to load their payload. After all, no astronaut has their spacecraft built around them!

The lesson works because the students have limited time, limited resources and are forced to creatively apply their knowledge to a tangible problem.

Water bombs, ready to go!

Building bridges using craft materials

Who can make their bridge hold the most weight? I've seen this done using newspaper and sticky tape, spaghetti and foam pellets, Lego® and more. There are multiple ways to build bridges, with the central learning being that students investigate different truss arrangements for strength. It's best to limit the students' resources, especially the sticky tape or glue, as otherwise students are tempted to set up their bridges with so much adhesive that they're not actually testing their bridge designs at all, rather, just how much they welded their structure together. For weight, we find that simply using school textbooks is more than enough to test the bridge truss arrangements to destruction. For those wanting better materials, see if you can use hot glue guns and wooden 'paddle pop' sticks (for Year 5 and 6 students, wearing safety glasses at a minimum and possibly gloves too).

Tower challenges

1. Marshmallow and spaghetti tower

The marshmallow and spaghetti tower build is lots of fun and teaches kids the value of triangles in engineering. Hand out equal amounts of marshmallows and spaghetti pieces to each group and give the students a time limit in which to create the highest tower. Students often have trouble with creating the base until they realise that a repeating pyramid pattern is the simplest solution. Once the challenge is finished, you can get the whole class involved in making one giant structure, especially if you get each group of students to make repeating pyramids that can be placed together (an example of streamlining a manufacturing process). The only drawback of this STEAM activity is that the spaghetti does snap quite easily, which means you need to have spare packets on hand and allocate some time for cleaning up. For this reason, the activity is best run on wooden floors or concrete, as the spaghetti pieces can be difficult to pick up off thick carpet or grassed areas.

Will this marshmallow tower survive?

2. Highest cup tower

Plastic cup tower building challenges are lots of fun and very easy to run. Usually, this challenge involves giving students 100 cups each and a time limit. Generally, the students start building pyramid-shaped structures. For an interesting variant, see if the students can create an inverted tower instead, whereby they start with one cup on the first level, then two cups for the second level, three cups for the third level and so on. The students will have to concentrate, as they will need to be very careful about the centre of gravity the entire time. To make it easier, you can include paper plates as part of the materials list.

Inverted cup tower in progress!

3. Straw tower

Given some straws, sticky tape, scissors and a time limit, the students have to build the highest tower. To make it interesting, you can make it a requirement that the constructed tower also has to support a weight, such as a wooden block. Often, the limiting resource is the amount of sticky tape the students use; however, kids can get around this by connecting the straws together for the upright columns. This is a fun STEAM challenge which can be run at a moment's notice.

Can the tower handle the wooden block's weight?

Boat challenges

1. Boat building challenge

This is an engaging STEAM challenge based on displacement and buoyancy. All you need to do is give the students a set amount of aluminium foil, a bucket of water and some metal nuts. In a limited amount of time, the students need to create the boat that can hold the most amount of weight. Younger students often struggle, because they design the boat to be streamlined rather than creating a barge shape. Students also often sink their boat due to placing all their metal weights to one side of the boat. In a discussion with your students, you can mention the importance of the Plimsoll line on freighters, which is used to measure the effect of the weight of cargo on the submersion of the ship and, thereby, how seaworthy the ship might be in rough sea conditions.

How many nuts can the boat float?

Balloon-powered boats

Boat building STEAM challenges are fun and this one gets the students going! Given a balloon, a plastic food container, sticky tape and straws, the students can fairly quickly put together a boat that will move in water. To make it more challenging, it's worth introducing this as a race, whereby you give them multiple balloons along with materials that could be used to make the boat more streamlined. As always, time should be a limiting factor.

Balloon boat ready for action!

There are so many options for running engineering challenges in your classroom. Other teaching ideas include constructing a cantilever to support a marble out from a desk, raising a basketball up using newspaper and sticky tape, or breaking out the Lego® to make a Rube Goldberg machine out of levers, pulleys and gears that knocks a Lego® figure into a glass of water. Engagement all comes down to the backstory you present the students.

Having these activities on hand as an extra extension activity makes for highly motivated kids and very little downtime. As a life skill, kids benefit highly from learning to follow up on statements such as 'I wonder if ...' and they definitely enjoy trying to solve a problem in a supportive yet competitive environment. The best bit is that having the materials on hand in your classroom won't break the bank, and the lessons can be set up in very little time. They have tremendous potential to expand your students' confidence in problem solving. On top of this, the creative nature of a STEAM activity allows your students to take risks and explore their own imagination and curiosity. The more the students get exposed to design challenges the better, as it can only result in your learners becoming more independent and self-directed – exactly the kind of outcome you want in your students before they leave school!

Once your students have cut their teeth on these craft-based engineering challenges, they will most likely want to try more complex problems and builds. If you've got the school executive behind you, it might be worth finding a spare classroom in which to create a fully-fledged Maker space.

Establish a Maker space and join the movement!

'Creativity is putting your imagination to work, and it's produced the most extraordinary results in human culture.'

— Ken Robinson

- How can you find ways for your students to apply their scientific knowledge?

- Can you let your hair down too?

I love the Maker Movement. The popularity of Maker spaces, Maker fairs, Maker nights and more has grown and grown from a grassroots movement into a worldwide phenomenon that brings Science, Technology, Engineering, Art and Mathematics together, enabling students to tinker and explore new ways of using materials and tech creatively. Using open source software along with materials from electronics and hardware stores, or even second-hand materials that can be re-purposed, the Maker Movement is fuelled by an inquisitive breed of kids and adults intent on creating DIY projects that aim to solve problems that commercial products either don't address or for which they are way over budget. Often, it's not even that; it can mean creating something just for the sheer pleasure of finding out: 'Can this be done?' Plus, it's fun to boot! So the question is: Can teachers create their own Maker labs in their school? What are the barriers and what can you do to overcome them?

Many teachers will instantly recognise the constructivist roots in this blend of tech hardware and hacker culture. As we've seen, kids learn best with their hands (and so do adults for that matter). While it's great to learn how science works from a theoretical perspective, it's only when we get our hands dirty that we explore and make connections with our world. Of course, as we've covered, teaching the scientific method is incredibly important when it comes to teaching science in school. However, running the classic science experiments completely by the book doesn't address science's hidden partners ... creativity and imagination.

Diving submersible at the Great Lakes Science Centre, OH

133

The classic line is this: We are all makers. We all like to create things. We all like to build things. We all like to personalise things. The Maker Movement just provides an avenue and almost a permission to act on these fundamental drivers of human endeavour.

As the pedigree of the Maker Movement is aligned directly with working with community, creating a Maker space in your school could bring highly engaged adults and kids together in a learning environment that they'll care for. The advantages of establishing such a venture at your school could go well beyond student STEM learning outcomes; perhaps it could be another way for your school to be a focal point in your locality. Plus, it might help drive both departmental and media interest in your initiative! If you're raring to go but you know it might be a bit harder to convince your team, maybe it would be worth playing Dale Dougherty's TED talk 'We Are All Makers' for inspiration at your next school staff meeting.

 Check out Dale Dougherty's TED Talk here! https://www.ted.com/talks/dale_dougherty_we_are_makers

Creating a Maker space in your school would go a long way to addressing the curiosity and inquisitiveness in your students. Of course, you will need to do your due diligence in terms of safety and child protection prior to establishing a Maker space, but if you have a spare room in your school, what have you got to lose? You could even use this as way of partnering with your local high school.

So what's actually involved? Maker labs are effectively a natural extension of a science club with an engineering twist. The cost of materials is entirely dependent on what you want to achieve with your Maker space. I've seen construction materials consist of cardboard to recycled electronics through to complete pre-packaged start-up kits to get you going. It comes down to how much time and money you wish to spend and, more importantly, motivation! This might even be an opportunity for students to learn how to prepare and implement a business case. Could you help them market it to the community and industry leaders for financial support or sponsorship?

Below is a list of some of the commonly used materials that go beyond simple craft materials; this is certainly just the tip of the iceberg:

- 3D printers and scanners.

- DIY electronics and old phones.

- Arduino & Raspberry Pi.

- Open source code and programs such as Scratch by MIT or searchable platforms such as Sourceforge & GitHub.

While it would be great to have the fancy support materials like 3D printers and so on, in reality, the materials don't have to be prohibitively expensive. All you have to do is consider how to re-purpose materials around you.

Are you wondering what other support resources might be available? Well, apart from the curiosity and inventiveness of your fellow amateur STEAM enthusiasts, there is an entire branch of media and associated businesses that have developed as a natural progression of the Maker Movement, including Makezine.com, Makers Empire and more. You can even check out a feature presentation called *Maker: A documentary on the maker movement.*

Worldwide, science and technology museums have been rapidly establishing Maker spaces in response to the growing demand. On my Churchill Fellowship, I was privileged to visit the New York Hall of Science, where a Maker lab is placed directly in the middle of the main auditorium to encourage visitors to come and work with odds and ends to complete engineering challenges in their own way. The kids were completely engrossed in their tasks, so much so that it was clear that one of the parents was struggling to get their kids to move on to other exhibits!

nySci Maker Space

If you would rather plan a class trip to an established Maker space, there are quite a few options to choose from. Everything from a local Maker fair in Australia to global Maker campaigns can be accessed with just a click of a button, and are worth looking into.

Regardless of what your position is on establishing such a room at your school, it's difficult to ignore the potential for these informal activities to positively impact learning. Even if you just bring one component of the Maker Movement into your classroom, you'll go a long way to inspiring your students to think, create, explore and apply knowledge. Isn't that what we do as educators?

ACTION POINTS

- Create some interesting scenarios when planning out your lessons. The students will love you for it and your classroom could quickly become a showcase for others to emulate. Can you take this a step further and link classes via video conference to simulate a mission control / spacecraft mission or similar?

- Look for the historical context and teach about when discoveries were made and why. What made the scientist begin to study the topic? What were the implications of the discovery? Was there any opposition to what was found out? It's the intrigue and stories that help make the content much richer, and the students will get more meaning out of the topic. Look up Big History to see how this all ties together.

- Plan to take your students into the great outdoors! Don't just make it a nature walk. Instead, have the students look around and evaluate what they see. Why are specific plants growing there? What caused the surrounding topography? Is there any impact from human disturbance? What evidence of animals is there? Just make sure that you've got the right permissions and safety documentation in place.

- Make arts and crafts a problem-solving activity. All manner of topics in the physical sciences can be taught using simple materials you can find hanging around your classroom. Students love competitions too; maybe this is the time that student groups can earn those gold stars!

- Consider creating a Maker space in a spare classroom. Start out small with craft materials and some hand tools and slowly build up from there. With some input from interested parents and maybe some funding, you might quickly find that the space you create is humming with creativity!

Notes

4. Leveraging technology

'Technology is just a tool. In terms of getting the kids working together and motivating them, the teacher is the most important.

— Bill Gates

ALL THE TALK about technology used in Maker labs brings us neatly to this chapter ... technology in the science classroom! You'd have to have been living under a rock not to see the massive change that's happened in the last twenty years when it comes to access to digital knowledge and machines. It doesn't cost that much to integrate technology into the classroom and the benefits of utilising tech with students are too big to ignore.

Of course, the issue with including technology in a book like this is that it may quickly make this book dated. Yes, it's a drawback of this fast-paced world, but I'd rather address some of the options open to you than pretend they don't exist! It will be interesting to see what endures.

Science apps for the classroom: Devices at the ready!

'If we teach today's students as we taught yesterday's, we rob them of tomorrow.'

— John Dewey

- Can you turn the explosion of smart devices throughout popular culture to your advantage in your classroom?

- Students will use smart phones whether you like it or not. If you can't beat them, can you join them?

With today's technology, there are so many options for enriching your students' classroom experience. One of the quickest ways to make it happen is by integrating science apps into your lessons. With the big push for BYOD (bring your own device) in schools, this is a way to get students actively involved in learning STEM by taking advantage of the digital environment that they've grown up with.

A quick search of the iTunes® or Google Play® education sections will turn up a plethora of apps, so many that it can be confusing to choose which to try out first! Regardless of which science apps are for you, there are some things to take into consideration, mainly around the practical side of lesson planning:

- Which science apps will produce the learning result you're looking for? You need to do your homework here, as you can download an app that promises a lot but comes up short in delivery. Look for the star rating and the number of times it has had feedback (the more the better).

- How much time is needed for the students to learn to operate the app? This is a significant factor given the short amount of time you have for science per week in the curriculum you have to teach.

- What is the size of the app? There's no point planning to download an app if you have no more room on your device!

- How will this affect your teaching? Will the students work in groups, individually or be guided by you via an interactive whiteboard? What is the learning sequence coming into the lesson, during the lesson and after the lesson? In other words, how does the science app fit into your learning sequence?

- What support materials do you need? Do you have the adapters to get an iPad® or Android® device onto your data projector? These don't cost that much, but it's worth allowing for this.

- Is the science app available on different platforms? This is especially important if you're planning on the BYOD approach, as students will have all manner of devices.

- Is your school's bandwidth good enough to support the app download during the lesson?

Finally, if you go with only downloading free science apps, you'll have to put up with in-app purchasing advertisements. Most of the best science apps have locked areas requiring you to purchase the full version. Perhaps consider downloading the free version to get used to the app interface and then, once you're comfortable, purchase the full version. After all, you do get what you pay for and the developers will only keep producing science apps if it's viable for them in the long run.

If this area is something you'd like to explore further, it could be worth joining your State's educational technology group (for example, ICT Educators NSW), or you could join the International Society for Technology in Education where you'll find the Mobile Learning Network and the STEM Network very useful!

With some planning, you'll find that using science apps will help invigorate your classroom and allow you to address some of the technology learning outcomes required in today's science syllabus. Besides the learning, the kids love it!

In the reference section at the end of this book is a short selection of science apps we've found to be useful in the science classroom. Some of these we've used a lot at Fizzics, while others we only discovered recently. As mentioned before, technology changes quickly, so these apps could become out-of-date very soon … but at least it's a starting point!

Another way to get students into technology is to spend some time learning how to code with robots. This is very useful when combined with a Maker space. Robotics combines maths, science and engineering in a spatial environment that students love!

Robotics: Teaching problem -solving in a digital environment

'In the twenty-first century, the robot will take the place which slave labour occupied in ancient civilisation.'

—Nikola Tesla

- Have you heard the major call in countries around the world for students to learn how to code?

- Can you find a robotics platform that will suit your students?

The call for learning to code has come loud and clear from across the globe as governments and businesses recognise the value in students understanding how computers actually work. This does not mean that these authorities envisage that the future workforce will have to know how to 'build a website' or 'control a robot'; the actual outcome of learning to code is that exposing students to the rigours of computational thinking will allow them to broaden their mathematical ability, logical reasoning skills and critical skills in any occupation and even leisure time!

There are a variety of free coding platforms out there:

- **Scratch:** https://scratch.mit.edu/

- **Code Academy:** https://www.codecademy.com/

- **Code.org:** https://code.org/learn

- **Google® Computer Science First:** https://www.cs-first.com/

It would be remiss of me not to mention the low-cost computing platform Raspberry Pi at this point (https://www.raspberrypi.org/education/), as these programmable systems can be used in a variety of Maker space challenges for older students. I'm also a big fan of the initiatives such as Code Like a Girl, apptEDUde, Code the Future, Code School, Code Camp and more (in fact, there are literally *hundreds* of places that have sprung up in the past few years).

Still, one of the most captivating ways of teaching coding is where you deploy robotics in the classroom. Student imagination tends to run wild once you give them the opportunity to control a robot, and you can find quite quickly that you can sneak in some advanced mathematics. Why? The positive reinforcement of the robot working as the students solve problems will have the kids more than happy to push their understanding of maths even further. Even if your school is short on cash, you can still teach the principles of robotics without a robot in your room!

Robotics lesson without the use of a robot – a great intro at all levels

Primary kids often think that the average robot can be extraordinarily intelligent and expect that robots can think for themselves with very little input from humans. After watching movies, cartoons and more, who could blame them for thinking this? A great way to show the difficulties in programming robots is through asking student to give you commands as if you were a robot in the room. The trick of this lesson is to be incredibly pedantic; insist on requiring each and every tiny bit of information to allow you, as the robot, to achieve the goal ... the students will find this game highly amusing / frustrating / engaging as they try to grapple with the problem of describing how exactly you move your legs to move, how you use your hands to pick things up and more.

1. Stand at one side of the room and ask the students to give you a series of instructions for you to simply walk to a table and pick up a bunch of grapes. Be sure to say, 'I will do *exactly* what you ask me to do.' Make sure you include some over-acting, complete with robotic noises as you complete each instruction, to make this really fun!

2. The students will almost certainly ask you to take a certain number of steps towards the table. Simply answer, 'Does not compute. How do I move my legs?' in a robotic voice.

3. At this point, the students will look puzzled, but then will try again by telling you to 'raise your leg'. Quickly reply, 'Does not compute. Which leg should I raise?' The kids will then respond either left or right, to which you could respond, 'How much do I raise my leg?' Once the kids specify a specific leg and height, using some balance, you could raise your leg from the hip while keeping your entire leg straight. The kids will then realise very quickly that you require them to describe each joint movement! They'll now ask you to more or less relax your leg so that foot drops and the upper leg still is raised.

4. So, you're now standing on one leg with a knee up in the air! The students will now have some difficulty in trying to describe how you actually take a step, whereby a number questions could be tested:

   • Do you push with your foot at some point during a step?

   • How much do you lean forward during a step? Is this even necessary?

- How do we describe pushing backward with your foot?

5. After some trial and error, the kids will, at some point, have achieved getting you to make your first step. Now is the time for you to call out in a robotic voice, 'Shall I save these instructions as my step routine for my right leg?' In unison, they'll all call out, 'Yes!'.

6. Now you've got a step routine for the right leg, get the kids to repeat this for the left leg too!

7. The kids will now be able to get you to move to the desk. They now have the considerable challenge of getting you to move your arm and hand into position to grab the grapes. Even when they get your hand into position to pick up the grapes, quite often, they'll tell you to simply 'close your hand' … causing you to close you hand too far and crush the grapes! Exasperation ensues but the students will have 100 per cent learned the lesson. They need to be very careful when giving instructions to an automaton!

Of course, you could be much fairer in a second challenge, whereby you give them a sheet with acceptable instructions that you, as a robot, can follow. This time, present an even harder challenge. The kids now have the 'programming language' and can guide you very quickly to achieve the result.

149

What you're doing is teaching them computational thinking, whereby students learn that most programming languages require you to specify very specific steps as a series of explicit instructions that can be understood. Computational thinking (also known as logical reasoning) is highly prized in modern society as the ability to break complex ideas into step-by-step instructions. This is a critical skill for engineers, scientists, managers, entrepreneurs and more. As an extension, you could now get students to 'control' each other via voice command for a number of tasks and have them note down just how many steps are needed to do basic tasks; e.g., bite an apple, brush your hair, kick a ball into a net and more.

Using our body to learn about robotics

Even the students' bodies can be used as an allegory to robots in the real world. Get students to work with their own bodies in an effort to get them to understand that electrical signals are sent to and from the brain (our own CPU) along nerve pathways (analogous to electronics found in robotics).

Here they can explore their five senses (an analogy to sensors on robots), they can feel their muscles contract and relax to varying degrees based on resistance (similar to increased energy requirements of motors under load) and more.

Lego® robotics

I love Lego® robotics! Also known as Lego® Mindstorms®, we've been using these robots at Fizzics for years to teach kids coding and logical reasoning. The beauty of the platform is that it is graphically based, whereby you can 'drag and drop' images into a programming sequence on your screen that represents snippets of code that control the robot.

Students intuitively understand it and can very quickly have their robot moving and sensing its environment without the need to learn the back-end code that powers it. You can also use Lego® robots to discuss robotics in general, in terms of why people have developed robots and the possible implications for future society.

EV3 Lego® Mindstorms® Robot ready for action!

What I've found when introducing Lego® Mindstorms® to a class for the first time is that it's best not to dwell too long on the intricacies of each and every function that the robot can perform. Rather, it is far better for the kids to get up and going quickly, briefly showing them the basic differences between the different motor block functions and setting them a movement task. This way, students can learn their own way – that is, by trial and error.

It can be interesting watching kids come to grips with the different motor blocks available. If it suits your students' learning style, they can choose to control the motors in several ways:

- Control the motors independently using the single motor block.

- Coordinate power between the motors using the 'move steering' block (most kids find this one intuitive).

- Vary power between the motors using the 'move tank' block. This more or less achieves the same function as the 'move steering' block, only the student can clearly see how changing the power levels on the different motors produces changes in direction (a bit like the skid-steering you see in a tank or an excavator).

Once the students can move the robot confidently forwards, backwards, left and right, it's time for some challenges that also act as a diagnostic on student ability. One of our favourite ways to test the students' newfound ability is via Lego® robot bowling!

As the kids gain confidence using the Lego® robot's basic motor and sensor functions, they can begin to learn how to get more control out of their robot using the loop and switching functions.

Loops

By placing a specific programming sequence into a loop – for example, 'move straight for one wheel rotation at power level sixty' – you can control the number of times that that sequence repeats. This can be quite powerful, as the students don't have to keep dragging and dropping repetitive lines of code to produce a movement; instead, they can choose what they want to repeat and drop it into the loop itself.

Switches

Also known as 'if' statements, switches allow you to have the robot use the sensors to gather information about the surrounding environment and then make a decision. A classic example of this is by using the touch sensor, whereby you set up the program so that if the touch sensor button is pressed, the robot moves forward for a set number of moves, and if it's not pressed, the robot won't move at all. 'If' statements are found throughout programming and are the basis of binary logic, which basically has the robot saying, 'If this thing occurs, I will do this ... and if this other thing occurs, I will do this instead'.

Knowing that the sensors produce results that are measurable means that you can begin to use these results as data outputs that control more functions of the robot. In one of the palette of programming blocks, you can find mathematical functions which allow you to manipulate the incoming data from a sensor and send this as an output that the motors or the display recognise.

So, having learned a bit about how the programming environment works in Lego® robotics, what are some of the challenges you could have the students try? Below are just some of the sorts of things you could get your students to trial:

- Clear a table of light obstacles using a basic programming sequence.

- Program the robot to stop on the sound of a loud hand-clap.

- Use a light sensor to detect light levels in the room.

- Use logic-based programming to make the robot follow a black line.

- Use a touch sensor or ultrasonic sensor to detect a solid wall.

- Program the robot to react to differing light levels in the room.

- Use a combination of motors and light detectors to hit a golf shot.

- Determine the correct force required to score an ice hockey goal.

- Make an Etcha-sketch® using the display and motors.

- Make your robot act like an animal (be scared of noise, hide from light, and more).

- Negotiate a maze of solid objects ... and more (it's only limited by time and imagination!).

There are a variety of Lego® robotics competitions that you could enter your school into as well. Two of the popular ones in Australia are First Lego League® and Robocup Junior®. With themed challenges, from dance offs to robotic soccer tournaments, these competitions provide an awesome goal for students to aim for. Not only are they an extension tool, they are also an opportunity for students to put themselves up against students from other schools in a fun learning environment. Of course, there is absolutely no reason you can't organise your own robotics competition within your school or between some friendly schools or libraries in your district. You could even form a school robotics club!

Students competing in the Robocup Junior® maze challenge

There are many ways of using Lego® robotics in your classroom, and you don't have to feel limited by the number of lessons you get with the software package. Try typing 'Lego® Mindstorms® lesson plans' into Google® and suddenly you'll be presented with all sorts of lessons, videos, books and more that can help you out. To be honest, though, sometimes, the best source of knowledge is the students themselves. Once they're truly engaged, you'll find them researching ideas to plug into their code to solve problems and they will be only too happy to share their knowledge. This is fantastic, as it is this kind of self-directed learning and collaborative culture that you want to encourage in your classroom in the first place! Soon, you'll find that the participating students will be willing to try more and more difficult activities. In no time at all, your classroom could be filled with creative thought and critical thinking that spills over into your other lessons. Not a bad outcome at all!

Perhaps Bee-bots is the robotics solution for you!

Got to love a Bee-bot®! They're such a well-designed robotics teaching tool for young students that they not only serve as an entry mechanism into more complex robotics but also, if used correctly, can be a versatile tool for teachers wanting to creatively teach classroom content. Yes, they look like a bee and young kids can get quite attached to them. You may initially feel limited by the set length of movements and ninety degree turns that the robot makes, but younger students don't seem to mind, and, if anything, the simple programming structure allows students to quickly master the movements, which means you can concentrate more on using the Bee-bot® as a vehicle to extend other learning areas.

Bee-bots reporting for duty!

There is also an extension of the Bee-bot® platform called a Blue-Bot®, a similar shaped robot which incorporates a Bluetooth connection that allows you to program the Blue-bot from a computer or tablet as well as the same buttons as per the Bee-bot®. Apart from the obvious use of being able to program via a software platform, the fact that the Blue-bot has a clear casing is fantastic as the kids can then see the inner working of the motors, lights, batteries and electronics.

Here are a number of fun ways that you can extend the use of Bee-bots:

Maze challenges

Setting a challenge for students is an easy way to get students motivated in any classroom! The basic grid pattern of forwards/backwards/left/right means that you can set a variety of mazes up using pens, insulation tape or simply coloured paper sheets on the floor. You can adapt a maze so that it is longer quite easily, and even use chairs within your room to create a tunnel.

- ## Bee-bot® dancing

Yep, you read that correctly! Not only do kids enjoy making the Bee-bots dance in a similar way to synchronised swimming, it also teaches teamwork and allows students to express their own creativity. Makes for a great end-of-task activity!

- ## Build cities, roads and terrain

Young students really get a kick out of creating their own built environment, so why not let them loose on Lego®, building blocks, road maps and more. You could have the students create an environment that the Bee-bot® has to navigate or create a challenge, such as making the Bee-bot® move from the hive to a flower and back again. Create a fun environment for student to explore.

159

- **Spell names**

  You can use Bee-bots to cover literacy too. Cut out some letters and stick them on the floor in a 15cm grid pattern. Students can then spell out their name or other objects. They could identify each letter by spinning on the spot at each letter required (you might need more than one Bee-bot® to finish off long names). Even with a spelling challenge, kids can't help but want to build on it too.

- **Mapping**

  If you get clear sheet of plastic and rule out a grid, you can place all sorts of maps under the grid for students to use. Try having the kids travel between cities, between suburbs or even a map of your school grounds. If you label the grid, you can get students to call out where to go, for instance, 'Move to B3 and then move to C6.'

- **Storytelling**

  I saw a brilliant use of Blue-bots and Bee-bots by my daughter's Year 1 teachers at Holy Cross Glenwood. The students had created a story about how a Blue-bot went for a walk in the woods along an enchanted path to learn how to do 45° degree turns. The students had constructed a path out of painted cardboard, cellophane and counting blocks with the inclusion of a bridge and tunnel. During the adventure along the Blue-bot had to shake vampire bats off his back (that had flown out of a treasure chest), pass through the dreaded 'hula hoop of fire' artwork, have a picnic with a bear and finally meet up with other Blue-bot and Bee-bot® friends on who would guide the Blue-bot to find the mystical instructions on how to turn 45° degrees. A cute story created and narrated by the year 1 students that students presented at school assembly using an uploaded YouTube® video which clearly showed their ability to move the robots as needed!

Despite the limited range of movements Bee-bots have, you can really use the grid pattern that they make to great effect. What you're really doing with the students is teaching them problem-solving, planning and sequential thinking in a spatial environment. The added bonus is that students learn that robots only work within the parameters you give them, certainly something they realise quickly when the Bee-bot® moves a different way to the one they expected! In the end, it's the context and usability that kids respond to, especially if you get creative with some challenges.

Another idea you could use with Bee-bots is to have students write 'programming code' for other students to then write down what they think will happen. The students can then enter into the robot their decoded instructions and then see if they got it right!

Classrooms without walls: Break through using video conferencing!

'Everything is theoretically impossible until it is done.'

— Robert A. Heinlein

- Video conferencing has few limits; where can you connect?

- What knowledge can you find beyond your borders?

One of the breakthroughs of the digital age is the ability to see and hear people via video conference. There is a huge number of software applications that allow your students to connect with a class or even a museum elsewhere in the world. The learning opportunities are immense for those prepared to try it out.

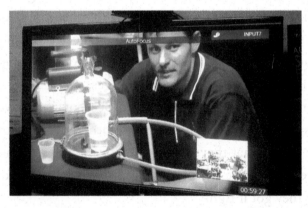

Demonstrating pressure decreases in a bell jar in a conference

One of the misconceptions about video conferencing is that it is a 'video' ... nothing could be further from the truth. A video conference is about interacting with the speakers – students asking questions and getting answers in real-time. It's about authentic lessons, and kids getting to run experiments in their room while being guided by a science expert – even if they're over 1,000km away or in another country! It has nothing to do with passive learning; it's about truly engaging experiences that students otherwise would never have access to due to geographic isolation or socioeconomic factors.

These days, there are web conferencing solutions that will allow your students access to educators with minimal hardware required. All you need is access to broadband internet, and a computer or smart device. You might be teaching in a school that has a proper video conference system in a connected classroom. If you're not sure, just ask your ICT support team or principal.

So, what are the sorts of student learning outcomes you might expect when you link your class with a science expert? Let's look at a couple of short case studies from my Churchill Fellowship tour of North American science museums:

Combining art and science using distance learning at the Cleveland Museum of Art

Museums can mix art and science in a video conference setting very easily, and students can get a lot out of it if you try it out. On visiting the Cleveland Museum of Art back in 2014, I found out how they have been connecting their in-house experts on art curation and preservation to students, discussing the chemical treatment needed to bring artwork to showroom quality. The team has also run video conferences for the visually impaired, whereby tactile materials such as pottery and Travertine are sent to students for the experts to then describe in detail. Such programs are highly rewarding for the museum presenters, the school teachers and participating students, as they encourage strong communication and descriptive skills. In fact, all three groups learn from each other as they share their own experiences and perspectives. In teacher professional development sessions, I've often argued that science doesn't have to be run by itself and that mixing it up produces richer meaning for students; it was great to see this in action! The animated introduction using Google Earth® as a world view zooming in to walking through the galleries was a nice touch for remote students, helping them get an understanding of where the museum is.

Arielle Levine presenting Knights, Castles & Kings on the chroma-key green screen

Using the on-hand artefacts, a document camera and a green screen chroma-key, the remote students can explore STEM outcomes such as the water cycle, rock cycle origami and tile tessellations, and still connect with the core artistic messages of the museum. One of my favourites is the simple machines lesson, in which the museum educator discusses medieval trebuchets and ballistae and has the connected students create their own mini-catapult!

Bringing remote students to the Arctic at the Alaska SeaLife Center

Set on the edge of town on the shores of Resurrection Bay, the Alaska Sealife Center specialises in cold water marine animals and is Alaska's only permanent marine mammal rehabilitation facility. Since commencing video conferencing in 2005, the centre has grown a reputation for strong content mixed with solid teaching pedagogy. They run highly informative distance education classes via video conference, in which students can experience real-time shots of seals, puffins and fish from the holding tanks, document camera shots of otter teeth and fur, see recorded footage of animal behaviours and more.

Using a shopping list of simple materials, the connected students can create models of seals in their room, and hear stories about sea kayaking with whales and living in remote Alaska.

Darin Trobaugh in Alaska presenting to a school in New York

The authenic way that the centre approaches distance education shows the real strength of

educational video conferencing: It is a two-way conversation, in real time, with a site that students may
never get to see in person.

And it is not only about a site's particular content;
the educators who run distance lessons have had
real experiences working in the museum/galleries/
zoo/aquarium that they are representing. There is a
depth of knowledge that students can access, and
it is as though the expert is right before them, with
the lesson changing to fit the questions of students
in dynamic classrooms around the globe.

There are so many ways you can use video conferencing to
discover new educational opportunities, from visiting world-
class museums and zoos to diving on the coral reef. Still, as a
teacher, you often get your training on how to use the newly
installed VC system and then are left to your own devices,
having to work out the intricacies of how to use it yourself.
If your school timetable means that you don't get to use the
system for a couple of months, you can quickly forget some
of things you need to know to make your experience much
more enjoyable (for both your students and the far-end site
working with you).

To help, here are a few tips that can help out when preparing for a virtual excursion, plus some simple hacks to
make you look like a pro:

Setting up your room

- Connect a computer to your system with a HDMI or AV connector. You'll then be able to share all sorts of content with the remote site. If you want, you can use an adapter to connect to your iPad so that you have a document camera with extra functionality.

- Have a remote mouse and keyboard at the table near you. That way, you can access photos and applications easily. It's useful to have a Google® page open so you can look things up on the fly as needed. Additionally, it can be helpful to have the batteries easily accessible so that if you run out of charge, you can quickly change them over.

- Consider the placement of your audio. Try to get the microphones as close to the middle of classroom as possible. Sometimes, this is not feasible; in that case, you can have someone next to the microphone to relay questions and answers coming from the back of the classroom. VC systems come with a variety of noise cancelling microphones. If you choose to run a web conference via your computer instead of a H.323 VC system, it's worth purchasing a USB noise-cancelling microphone, as the sound quality for the remote sites is far superior to that you usually get with the inbuilt microphone on your computer.

- Put a whiteboard right next to your TV or projector screen and list down the different schools attending (especially if you're leading the conference). That way, you can quickly glance at the list and know whom you're speaking with. It can really help if you have a magnetic button that a volunteer can quickly slide along so you know who spoke last!

Before the conference

- Set your camera presets before joining the conference. This means a view of your entire classroom, a view of your experiment table and perhaps three separate views of the left, middle and right-hand side of the classroom. It can help to have a whiteboard view as well, so you can quickly write things down to show the other site. How do you do this? Generally, if you zoom your camera to the view you need, you can hold down a number button on your VC remote and a preset will become stored for later use. Check your user manual for how to switch between views (Polycom is slightly different to Cisco, for instance).

- Learn how to toggle the various layout views from self-view and 'Brady bunch' view. I personally like the self-view option, as I can see exactly what the other sites are seeing, and I simply toggle back to active speaker view so I can see the other sites when they speak back to me. It certainly helps me see what's going on!

Close up view of some chemistry
demonstrations

versus

Multipoint view layout showing all schools
involved in a conference

- If there are windows in the room, close any drapes or blinds. Daylight is a variable light source and can conflict with interior room lighting. Try to avoid 'back-lighting', as you will come across as a shadowy figure with your faces hidden.

- When adjusting your camera, try to fill the screen as much as possible with people rather than with the table, chairs, walls, lights or the floor. People want to talk with people!

- For microphones not fixed to the ceiling or table, ensure they are at least one metre away from the video conference camera/endpoint and not near any other electronic equipment, otherwise audio will be severely affected (think audio screech).

- Have the experiment materials and volunteers ready in your room so that the conference can be more interactive. Your students' experience of video conferences is affected greatly by what they get to do. Imagine you are a student and have to sit by and watch another school do fun science experiments in their room without you getting to have a go, all because your teacher's plan is to run the experiments later. You won't have much fun and will certainly report to the teacher that you don't like distance learning as much. Give your students the best chance to get the most out of the conference.

- Prime your students so that they're ready to ask and receive questions during the conference. Some teachers get kids to research the content quite deeply and have a question written down so they don't forget. Just make sure that they don't spend the time worrying about their time to speak and rather engage in the general conversation that occurs in the virtual excursion.

During the conference

- Keep microphones muted until invited to speak by the presenter. Mute your microphone directly after speaking. There is always peripheral noise at schools and the microphones pick up everything. This is a big distraction to the presenter and other participating schools.

- Use the presets you set up earlier to show the other sites images of your students, your experiments and the classroom itself. It's great to have a view set up showing what's outside your window too – kids love to see outside as a peek into your world.

- Due to the nature of some schools' allocated broadband speed, there is a chance of reduced clarity in picture. If a 'dropout' occurs, simply redial into the conference.

After the conference

- Like any lesson, this is the time to consolidate learning by asking questions of your students about the content presented, and to run aligned follow-on lessons. The more you align a video conference with your standard learning sequence, the more valuable it becomes.

- As with all things, the more you use it, the more familiar you'll become with what you can do, and the better your experience of virtual excursions will be. Learning to run one of these systems can be fun if you contact a colleague in another school to try some test runs with you. Even better, get some of your own students to learn the controls. This means that you won't always have to rely on having to set it up yourself, plus the students will get extra technology learning outcomes and confidence using modern communications.

Using drones to teach science

'Imagination is the highest kit one can fly.'

—Lauren Bacall

- Can you give your students a new perspective on things?

- How can you use the camera's unique perspective to explain spatial concepts?

The advent of drones has been heralded a potential game-changer in certain industries. Quite a few readers will be well aware of Amazon®'s interest in creating delivery drones for their online shopping business via Amazon Prime Air®, and you might have even heard of Dominos® Pizza Delivery drone too (see below). The consumer market is booming with drones that are within budget limits for some families, and some kids in your classroom might very well be using drones on a regular basis with their older siblings, parents or extended family. So, let's look at some ways you can work with your students' interest in this technology to give them a new view of science from above!

Now, of course, in a primary school context, as the teacher, you should consider being the person in control of the drone in your school. While there could be students in your class who know how to use one, it's best that you take responsibility for where the drone flies and how high it goes. Just be aware that different jurisdictions have different rules, so it is worth checking what your local laws are in regard to using drones. Like always, you should complete a risk assessment on the activity too, so that the activity is well-run and safe. Now we've covered that, let's look at how drones could enrich your students' science experience!

As drones have cameras attached to them to help the remote person to control the device, you can use this bird's-eye view of the world to give students a unique perspective of science concepts while using a gadget they love. Here are just a few ideas that you could use to capture the imagination of students in a science education context using drones:

- As you can control the height at which the drone flies, you can effectively control how close the on-board camera is to any given object below; this means that you can fly lower to 'zoom in' and fly higher to 'zoom out'. This could be used as an analogy of a microscope focussing on a prepared slide! Why not have students pretend that the camera view is that of a microscope and get them to construct large versions of cell structures on the ground? You could also use the same idea to create the structure of a leaf cross-section, or perhaps maybe a neuron? No matter what you choose, you could then call out the structures you would like the students to 'zoom in on'. This might be a fun twist on the classic 'build me a model' lesson plan and, in reality, the students could create models of any object that you're interested in viewing.

- You could create a line of planets in our solar system, all marked out at the correct relative distances along the length of your school oval. This time, you could launch the drone from 'planet Earth' and then have the students measure how long it takes for the drone to reach the various 'planets' to land. This activity would be analogous to the time it would take to send a real spacecraft to objects in our solar system and is simply an extension of the classic solar system model lesson plan.

- Why not use the drone's video feed to map your local bushland area? By taking photo stills from the video taken, your students could create maps of areas of undisturbed bushland vs. areas where urban impact is apparent (e.g. from weed invasion). You could also use this to see if your students can identify vegetation structure types (i.e. woodland vs. grassland) or even if they can identify individual tree species based on the shape and colour of the foliage. This activity would work very well while taking your students on a nature walk, especially if they're interested in technology applications in the real world.

- One of the difficulties students have in science and mathematics is trying to visualise how quickly things magnify when you multiply by ten. They have to get their head around it later in high school when they do microbiology, but you can also experience this in primary school when you ask them to compare earthquake strengths recorded on the Richter scale. Why not create a simple demonstration of this by having the drone move upwards in 10x increments, i.e. 1cm, 10cm, 1 metre, 10 metres, etc. Of course, the drone will struggle to make the higher levels, but the students will get the idea very quickly!

- Why not use the video from your drone in your next video conference with a distant school? The partner school's students would love to see more about your school and a drone will bring a unique vantage point to the conference. You could have the students compare and contrast aerial photos taken of your school, which you can show via Google Maps®.

- Drones really bring a new perspective into the primary science teaching context. If you find that the cost of drones puts the technology out of reach, you could still use the zoom function on tablets as you view images to simulate changes in heights regardless. Google Earth® is also a powerful piece of software that simulates drones flying all over the world, and you can discuss the impact of camera resolution on image quality.

Some more quick ways to integrate technology into a primary science lesson

'Providing better computer science education in public schools to kids, and encouraging girls to participate, is the only way to rewrite stereotypes about tech and really break open the old-boys' club.'

— Ryan Holmes

- Technology is incredibly pervasive but becoming incredibly user-friendly; how can you exploit this in your classroom?

- If technology is not your friend, it will undoubtedly be that of your students; how can they help you?

Still after some ideas to integrate technology into your classroom? Here are twelve quick ways you can introduce more technology into your classroom that are fun and highly rewarding for students.

1. Create your own document camera. Link a smart device with your integrated whiteboard or TV and use the camera the way you would use an overhead projector. Why would you bother? This is a fantastic way to show the minute detail of objects to kids at the back of the classroom.

Using an iPad® as a document camera

2. Use digital microscopes and have the students draw and label diagrams of all manner of objects (think feathers, newspaper, flowers and so on).

Digital microscopes are great for biology and more!

3. Use QR codes for students to scan and find out more information. I've seen some great treasure hunts set up in combination with compasses this way! You can generate free QR codes via http://www.qrstuff.com/ and you can find QR scanners on both iTunes and Google Play®.

4. Set up a camera to take pictures every 0.1 seconds to measure velocity and acceleration. You can use this to look at gravity, passing cars or hitting a cricket ball! The better the device, the better the results. You can also use the timer taken on a video to achieve the same results by taking screen shots of paused video too.

5. Use the accelerometer in smart devices to plot a graph of student movement. We did this once on a jumping castle and in a Sumo suit one year ... there are free apps you can add to your phone for this.

Holly Kershaw bouncing around for Fizzics using her phone as a data logger!

6. Use your smart device as a data logger for showing kids sound waves on your screen. You can show sound waves quite easily and by playing pure notes from a guitar or any other musical device, you can then demonstrate that when you have a high sound you have shorter wave lengths (frequency), and when sounds are loud the sound waves are taller (amplitude). Again, you can find free apps for your device quite easily.

7. Have students film an experiment and blog about it through your school's preferred platform.

8. Students can create podcasts and load these onto your class blog (they can log their experiences, interview a subject expert, describe a school event … whatever!).

9. Have a BYOD day and challenge students to creatively use their device to solve a STEM question.

10. Create a Rube Goldberg machine, whereby you set up a ridiculously complex machine just to perform a simple task. You might remember the board game 'Mouse Trap', which was based on this.

11. Check out the virtual reality headsets that are quickly hitting the market, as well as the use of 360-degree cameras for filming. What landscapes can your

students explore on the internet or document for others?

12.   Perhaps augmented reality might be of interest to your students? Have a look at such apps as Aurasma®, Augment®, AR Dinopark® and more.

App in Focus: Pokemon Go®

Well, hasn't this app exploded across the world! One of the most viral game releases of all time, Pokemon Go® use in August 2016 surpassed even Twitter in its daily users, and it will be interesting how long users stay with it. A worldwide phenomenon, it has been launched by Niantic® in celebration of the twenty-year anniversary of the original Pokemon® card playing game in Japan. With so many kids, teens and adults using Pokemon® Go, surely this is an opportunity for teachers to use this highly addictive content in the classroom to engage students, especially those kids who don't readily engage with your standard science lessons.

If you aren't completely up to speed with what Pokemon Go® is all about, here's a quick overview:

• Pokemon Go® is an augmented reality app. What this means is that the app uses your smart device's GPS location services and camera to overlay images of Pokemon® cartoon characters into the viewing screen when you use the app.

- The Pokemon Go® characters are shown on your local map at a 'Poke Stop' and your job is to go and find them and catch them by shooting them with a 'Poke ball' to gain points. Each Pokemon® character is different in its look, height, weight, etc., which is recorded in a 'Pokedex'. The Pokemons could be in shops, parks, streets or more, and they flick in and out of existence within the game environment. Incredibly addictive and easy to use, the Pokemon Go® app has caused the strange sight of people flocking to locations where the augmented reality characters are, talking about what they've found and then rushing off to the next location!

- You can get Pokemon Go® on iTunes® and for Android® on Google Play®.

While this may not be everyone's cup of tea when it comes to online gaming, you can't deny that Pokemon Go® has certainly hit a nerve among the general public. As teachers, it's our job to keep up with the times and use any means necessary to encourage learning in the classroom, and Pokemon Go® could be yet another way you can do this. The following list provides just some Pokemon Go® ideas that could be used in a science lesson; some are analogies, some are more direct, but all could be good fun in your classroom with the right introduction and context in the real world. It'll be interesting to see what tactics work and what new lesson ideas emerge ... either way, it'll be a bit of fun!

1. Get students to find other augmented reality apps. For example, I remember using an augmented reality app during the 2011 Jurassic Science week for National Science Week.

2. You could spend time teaching data literacy. How many times do the Pokemon characters turn up on a map? What has the class collected so far? How much time did each student spend collecting the item? How far did each student walk on each Pokemon collecting mission? The students could create graphs, tables and surveys around these and more. The 'pokedex', which lists the attributes of each Pokemon character, could also be looked at – for example, you could use their weight and height to determine their Body Mass Index!

3. You could discuss the advent of Global Positioning Satellites and their application in real research. You could even discuss how your smart device's GPS has to correct for relativity due to the speed of the satellites above!

The ability to use maps is very important. In fact, your traffic predictions on your smart phone uses network theory to work out exactly how to get from point A to B in the quickest fashion. This also raises the question of how these apps use data storage and real-time tracking of devices to produce relatively accurate estimates of how long it will take to get a location based on traffic congestion.

These are just a few ideas for science class. What are yours? In the meantime, just be sure to give safety instructions to your students. There have already been several incidents where people have walked into moving traffic while chasing after Pokemon Go characters. Furthermore, there have been sad reports of robberies occurring in some secluded sites where the augmented characters have been. Basically, students should have an adult with them if they're young, participate in the game with friends and in well-frequented public places for safety, and should use common sense while using the app.

ACTION POINTS

- Put your smart device to good use and download some science apps! Consider buying a connector to hook your device up to your interactive whiteboard or data projector.

- Find a robotics platform that fits your budget and let students go for it!

- Start to use video conferencing to your advantage! You have access to so many resources from across the world if you try it, and your students will become worldly in the process. Check out the list of resources in the appendix of this book to get started.

- Start to use the sensors on smart phones to act as surrogate data loggers.

- Teach your students coding using free platforms like Scratch® by MIT.

- Look into virtual reality headsets and augmented reality apps too.

Notes

5. Exploring social media

'It's fine to have social media that connects us with old friends, but we need tools that help us discover new people as well.'

— Ethan Zuckerman

Having just looked at a few ways you can use technology in class, let's explore the social aspect of the online world, and how you can use it to focus students on primary science learning outcomes. It's very much the flavour of this generation, so it's certainly an area of students' lives that should be taken into consideration. As one of their greatest influences, you can help direct your students to become better producers of useful material to share worldwide.

Classroom science blogging

*'Blogging isn't about publishing as much as you can.
It's about publishing as smart as you can.'*

— Jon Morrow

- Can you encourage your students to recount their learnings in a language that is less formal and more personal?

- How can blogging teach communication skills around knowing your audience?

So you've gone to a massive effort to get together the best materials for a science lesson and your class is completely engaged and thoroughly on task ... why not document it? Even better, why not have the students document it themselves, and blog about their science experiments and subsequent discoveries? Students love to hear from other students. They like to share stories, behind-the-scenes video, funny anecdotes, and, above all, they love to have ownership over what they are doing in the classroom. There is no better way to enable this than to give them the keys to a supervised education blog that not only allows the students to creatively express what has been happening in your classroom, but also showcases your own work for other classes to emulate. In short, your everyday science lessons could become a source of inspiration for another school and/or your community!

Having students contribute to a shared classroom blog creates unity and shared purpose among your students. It allows them to become familiar with communicating their thoughts and ideas using technology. Most science experiments are naturally visual and by their very nature have a narrative that can be easily put into a blogging context. The actual process of blogging is quite easy and can be highly rewarding for both your students and your school. However, before you implement a classroom science blog, there are a few things you'll need to consider:

- What's its purpose? Is it simply a managed outlet for student expression or is it a component of an overall assessment your students need to complete?

- What is your school's policy on blogging? A quick word with the executive will save you a lot of hassle. Nearly every school will be happy to support you when you demonstrate how the student blog will be implemented, how the proposed content aligns with your curriculum and, finally, how you will manage the students and public throughout.

- Child protection issues: Can your student images and recordings be shown publicly on the internet? You need to make sure that parent permission slips are signed and filed away by the school before you embark on your first class post. When drafting up the parent permission slips, be sure to be clear that the purpose of the permission is to cover student images, videos, audio recordings and any other materials that can be used to identify your students. Just like any other parent permission slip, you should run this by your school's executive so they're happy with the content and are also aware of what you're embarking on.

- Management of comments: Sadly, there are people on the internet who have nothing better to do than bring you and your students down. Thankfully, you have a couple of easy ways to deal with this. You can simply disable public commenting and control the users by employing a password-protected blogging service like EduBlogs® (powered by Wordpress®), Weebly® for Education®, Wikispaces Classroom®, PBWorks® or similar.

- How much time do you have to dedicate to the student science blog project? Ideally, the students should be documenting photos and video throughout your planned science experiments so they can be saved on a central file for access during a computing lesson later.

- Finally, is there a target audience for your student science blog? Is it designed just so students can share the information between the classes in your own school or are you thinking more broadly to include the students' parents, other schools or the general public?

And that's it! Once you've addressed the above, you have the framework to quickly throw together a student guide, some science topics for students to focus upon, and perhaps even an assessment rubric as well. As always, what your students produce will be the direct result of the science stimulus items in your class, the types of technology you have available, and your students' proficiency in manipulating science materials and using digital technology.

Below are some ideas that you could use to kick off your classroom blog articles:

- A blow-by-blow account of a longitudinal science experiment. By this, I mean the sort of science experiment that requires regular measurements over a series of days or even weeks. The types of science experiment that are well suited to this include crystal growing experiments, plant growth experiments or simply measuring the weather. This type of science blog is effectively a diary, where you publicly record your class's results as well as their thoughts.

- A description of how to use a particular science apparatus in your classroom. The science blog article could be about a digital microscope the students have been using, how they're preparing for an astronomy night, or about their preparations to put 3D printing into your Maker lab. Ask students to make their writing as informative and interesting as possible, as the aim of this piece is to open a window into their world.

- A record of a special school science visit, a science fair or a bush walk you took your students on recently. We come across quite a lot of class blogs where teachers and students have taken photos and videos of our science demonstrations, described them in detail and then used them as seeds for further investigation.

Importantly, the length of the blog doesn't have to be set in stone. Some days, you might find that students can only produce 100 to 200 words, and other days, the blog could be double or triple that! As your students get used to writing their blog, you should find the length of their submissions increasing, as well as the writing quality improving. The requirement to include variables and fair testing, and structuring their experiments using the scientific method, should make student blogging more structured and coherent. Why not have the blog address what you'd expect to read in a science report (aim, method, results and so on), and then allow them to write their own thoughts in a discussion at the end? As students progress through school, they should be learning to write a scientific report anyway; the blog could easily become a way of re-purposing this work so that it can be digested by other readers.

Try to include images and video as much as possible. Not only does this help document the science experiments in your classroom, but it also makes the content your students are producing much more engaging and inviting to explore.

Upload videos to YouTube®, Vimeo® or Vine (which I'll discuss in more detail below). You should find a 'share' button on the video platform, where you can get the code to embed the video on your blog.

Include lots of hyperlinks to information throughout your blog, as well as educational content you can find on other websites as well. This is a lesson in itself for students; they will need to learn to be selective when it comes to which website they feel most correctly and visually represents their science topic at the appropriate reading age. As much as I love Wikipedia, perhaps it might be worth looking at websites belonging to renowned authorities in a given area (the World Health Organisation makes complete sense if your students are blogging about medicine, for example). It's also worth considering timing the publishing of your student blog article to coincide with national or global science events.

Once you are producing regular educational content, on a weekly or fortnightly basis, and weaving the production into your curriculum as part of your routine, you'll find it gets easier and easier. Why not appoint a student as a head blogger each week? This head blogger could oversee collating your week's science activities into one folder so that every student can easily access the imagery and videos to quickly put together their blog article at the end of the week. Quite a number of schools now include blogs as part of the homework for students to complete, which is not a bad way of having students

195

reflect on what has happened in your class and show their parents what they've been learning in science lately.

Getting a classroom blog up and running doesn't have to be onerous. These days, most software platforms are very user friendly and have a large amount of resources. There are also forums where you can chat with other users and gain support and guidance. Once up and running, you'll find that the process is highly rewarding, and your classroom blog will become yet another student science engagement tool you can employ. The bottom line: Students will love it and so will you!

Blogging is just one step into the social media world. We all know about giants such as Facebook and Twitter, so why not look into some of the other lesser-known platforms? These can be game changers in the hands of primary teachers who use them wisely. First up, Pinterest®!

Pinterest® for education: It's a hive mind of ideas!

'Pinterest® has become popular because of passion, not profit.'

— Scott Stratten

- What if you never lost a lesson idea again?

- How can you find out what other primary science educators have been doing around the globe?

Ever been planning a science lesson but, for the life of you, you cannot remember that experiment website you need? Pinterest® is most definitely the education app for you! The platform is a free virtual corkboard where you can sort and save all those science experiment procedures and the background information that you need as a practising teacher for quick access. As Pinterest® is almost a completely visual medium filled with photos and videos, it works very effectively, as you can quickly scan through your boards for a picture that reminds you of the science lesson you want to teach ... and it often comes with a link directly to the site where it came from.

How can you use Pinterest® as a teacher? Well, when you're looking through the internet and stumble across a webpage that you want to save, you can simply 'pin' an image from that website onto one of your boards for you to access later. Because you can sort your 'pins' onto different boards and give them names, e.g. 'sound experiments', you can very quickly organise hundreds if not thousands of lessons in a way that makes sense to you.

What does it look like? Imagine a series of corkboards, each with a title representing a particular area of interest, such as 'science experiments' or 'classroom management'. Now think of each pin as a Post-It® note stuck onto this corkboard, only in this case, the pin has an image, a description and, in most cases, a link to a website resource. Using the site is incredibly easy. When you click on any of the boards listed on someone's Pinterest® profile, you then get to scroll through the assorted pins that have been placed on that board. Not only does each pin contain useful information you can use, you can also send the pin to a colleague as an email.

It would be highly useful, as a classroom teacher, to begin to curate your own Pinterest® profile filled with pins from websites that you care about. It's like a mind map of ideas, thoughts and resources that you control, so it becomes a highly useful way to store your best lessons and activities for future use. This means you will spend less time trawling through loads of text on websites trying to work out what you need or, worse, being stuck for ideas on a lesson. All you have to do is visit the boards you made previously! The more you use it, the more powerful it becomes.

Like any social media site, you can visit other people's Pinterest® profiles to see what they've been collating. If you like what they've been curating, you can choose to follow their entire Pinterest® profile or simply follow one or more of the boards they've made. More importantly, you can copy their individual pins onto your own boards, which means you can very quickly build up a resource list for your references. As you follow more and more educational profiles on Pinterest®, your home page will become highly varied as recent pins being uploaded become visible and available to keep as well. This is very handy and can work as a way of expanding your knowledge in a topic area very quickly.

How do you begin to add pins to your own Pinterest® boards? First, you need an account and then to create boards on the site and name them. Once you've named at least one Pinterest® board, you'll need to click on 'create a pin' to add your first pin. Now, you have a couple of easy ways to create a pin: You can either copy and paste a webpage URL, whereby Pinterest® than trawls that page for images to pin, or you can upload an image from your device. You just need to describe the pin and you're done!

Despite this all being easy, it is even easier if you install the Pinterest® button on your browser – once you do this, it will be a simple matter of clicking the little Pinterest® logo when you find an image you like on the web. You can then place that image straight onto one of your Pinterest® boards. You can also install the mobile app on your smart phone through iTunes® or Google Play®.

The reality is that, over time, you will be exposed to more and more information on new ways to teach science, new science experiment procedures, developing scientific research and more. There comes a point where you risk simply forgetting all those great articles you've seen, and you get that frustrating 'it's on the tip of my tongue' syndrome when lesson planning! The good news is that there are software tools out there that can help you, and Pinterest® is a fantastic solution that can be quickly adapted by the busy science educator. You might want to check out the Fizzics Education Pinterest® profile when you've signed up yourself, for plenty of ideas to put into practice.

Topics for boards we've created and you could too!

- Early learning science
- Australian flora
- Science fair projects
- Science and art
- Plant science
- Magnetism
- Historical scientific artefacts
- Science songs
- Innovative science demonstrations
- Science news
- Geeky stuff
- Great scientists
- Museums around the world
- Environment
- Human pre-history
- Science YouTube® videos
- Explorers
- Science education humour

- Forces & friction
- Geography   Simple machines
- Food science
- Floating & sinking
- Microscopy Fun science experiments
- Classroom science craft
- Fun science toys
- Engineering projects
- Science teaching ideas
- Science apps
- Science books
- Science curriculum
- Slow motion science
- Strange creatures
- Education infographics
- Agriculture
- Hot & cold experiments

200

- Science events
- Beautiful science photos
- School science trips Christmas science experiments
- Science classes by video conference
- Teaching maths
- Vacation care science activities
- Science fiction
- Riddles, trivia & conundrums
- Archaeology
- Technology from left field
- Easter science
- Pressure experiments
- Energy transfers Fossils
- Teaching sound & light
- School science shows & incursions
- Science radio & podcasts
- Astronomy & space
- e-Learning
- Science websites

Now let's have a look at the value of getting students to document their lessons in short form videos... 15 seconds is all they need!

The value of short-form videos for teaching

'I guess I have a short attention span! I'm interested in new worlds, new universes, new challenges.'

– Alfonso Cuaron

- How do you teach students the art of being succinct?

- Can you give your students a creative challenge like no other?

With so many video-creation and sharing apps around these days, documenting learning with a smart phone or tablet is easy. Why not make it a bit more challenging by getting the students to create short-form science videos of less than fifteen seconds?

At a glance, getting students to create micro-videos might seem to have little teaching value, however, if you look a little closer, it could be the creative opportunity that your students are looking for. If you browse YouTube®, Instagram® or Vimeo®, you'll find millions of short films. Some are real gems. With a little bit of effort, people have used a very small amount of screen time to create useful science content that others can benefit from. Perhaps your students could try this out in the classroom. Apart from engaging students with short attention spans, having students create micro-videos provides an avenue for your students to get creatively engaged with your class's science activities. They often carry a smart phone in their pocket anyway, so why not let them film something great in your classroom and then get them to send the videos on to you to upload on your class's blog or social media platforms?

Taking this a step further, perhaps you could create some sort of 'science video expo' for your school and set challenges, with prizes for the students who can create the most informative and creative science video! This has been done in the past on YouTube® with longer videos to great effect, but I really love how the constraint of creating a video that last less than fifteen seconds forces kids to think very carefully about what elements are most important to include. To help make each student's video more meaningful, it would be worth showing your class how to quickly use video editing software to annotate what is happening during the clip and rapidly splice even smaller video clips together. Running a science video challenge could be a learning outcome in itself, as well as resulting in a showcase of your collective students' talents!

It is not the aim that the students create a Hollywood blockbuster. However, while most short videos are not necessarily the most stunning videos out there (we are teachers and students, after all), we've found that the very short timeframe forces students to get to the point quickly and hone in on the good stuff. In fact, you might be amazed at how quickly students will pick this lesson up and run with it. As long as the science is solid and it doesn't distract from the main focus of your lesson, your students can only benefit from giving this a try.

The most popular and well-known social tool for sharing short-form videos was Vine®. This allowed the upload of videos of just over 6.5 seconds in length, which would play in a constant loop. Unfortunately, this platform was disbanded in early 2017, which means you'll have to use other social media sites to upload the students' final video files: The most popular choice is Instagram®. With Instagram®, you can upload a video ranging from three to sixty seconds long (at the time of writing). Also, Instagram® offers another app called Boomerang, which allows you to combine your pictures to make one-second loop videos (almost like a rolling .gif file). Of course, you could also upload these to YouTube® or Vimeo®, both of which allow you to create a private channel to keep the videos away from the public domain if needed (you could use digital portfolio tools such as Seesaw® to keep the videos away from the public eye too).

It is important to note that if you decide to have students upload content onto social media, you will have to set the usual boundaries, expectations and safety guidelines, as with any publicly accessible website per your school's policies. Still, it's a fun and creative engagement tool that doesn't take too much out of your day. And you might find that you have a hidden talent for directing short films!

There are hundreds of social media platforms that you can explore and use within your classroom. While I don't want to go through all of them in detail, I now want to highlight an organisational app that has a social aspect to it. On Trello®, students, fellow teachers and even interested parents can work together. It's an incredibly efficient tool when it comes to project-based learning and more.

Trello®: A social media solution to classroom organisation and lesson planning

'There is incredible potential for digital technology in and beyond the classroom, but it is vital to rethink how learning is organised if we are to reap the rewards.'

— Geoff Mulgan

- Is there opportunity for you to get more organised through cloud collaboration?

- How many ways could Trello® boost your classroom efficiency?

I couldn't agree more with the quote above; in some ways, this digital century has overwhelmed us with information. One of the best ways to handle this is to combine the power of Pinterest® with the organisational app Trello®. This is the kind of app that allows you to keep track of tasks and their due dates, assign these tasks to students or colleagues, and provide resource links to further support material. The best bit is that Trello® is collaborative, which means you can invite all of your students onto the platform. Here, they can then clearly see what is needed, and learn not only to be organised but also to work together as a team.

The arrangement of Trello® is somewhat reminiscent of Pinterest®, in that you can produce a series of boards that describe a task. However, unlike Pinterest, you have the ability produce checklists, due dates and more. These days, there are so many little tasks a busy classroom teacher needs to keep track of that it can be overwhelming, especially for a new teacher on their first placement ... so why not give Trello® a go and see what you think? In fact, let's make this section of the book interactive – go and download the free app right now. If used effectively, in line with the guidance below, it will make your lesson preparation and delivery so much more efficient.

Deployment of Trello® in the classroom couldn't be easier. It is available on desktop computers, as well as on iOS® and Android® devices via a downloadable app. After setting up a free account, you can quickly create your first board.

The home page of Trello® once you first set up
your account. In this case I've named the first
board 'My science classroom'

From there, you can start adding lists that you want
students to contribute to (for example, weekly student tasks,
weekly newsletters, student projects and so on).

These lists need not be tasks in themselves; columns can
be labelled with each student's name, and you can assign
tasks directly to each student instead.

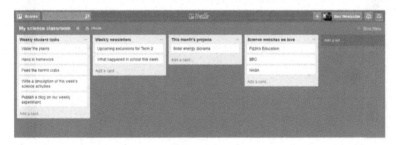

The Trello® board being populated by lists: Weekly student
tasks, weekly newsletters, projects, etc.

Once the lists have been created, you can start adding
'cards' to the lists that refer to the actual things that need to

occur (for example, research solar energy, gather materials for the class project, hand in this week's homework, water the classroom plants and so on).

If you click on any card, you can assign the task to a group of students, add a due date, add a checklist, link to an attachment and more. These cards can be dragged across to any list. Once the students get the hang of Trello®, they'll be soon doing this themselves!

Individual card on Trello® – note how clear it is for students to know what is needed

For security and child protection, you can set privacy settings to ensure only invited members have access. And you can grant access to different groups for different boards, while keeping other boards completely private.

Once you're set up, you might be interested in some ideas for using the app across the entire school, as well as in the classroom:

- **Lesson planning and collaborative delivery**

  Whether you work by yourself or in a team-teaching environment, you still need to plan what is coming up each semester. Planning your lessons is very easy using Trello®, as you can create a list for each teaching week and then assign cards for each lesson. Under each lesson card, you can then note teaching resources needed, stipulate whether you need to book the computer room, add lesson objectives, attach worksheets, link to supporting websites and more. If you're team teaching, you can assign tasks to other members of the board so that you can coordinate more effectively, checking the board to see who is doing what and by when. You can set up email notifications that alert you when a member has added a card to a list, updated a checklist or assigned you something to do. Another powerful way to use Trello® is to write down the curriculum requirements as individual items on a checklist and tick them off as you achieve them.

- **Student collaborative projects**

  Project-based learning is a major focus these days and getting students to work together can be a challenge. Some of the issues can be quickly sorted, however, if they have a central point to collate their thoughts and begin to plan their activities. Using Trello®, the students can clearly define the task steps and become more effective at planning. More importantly, they can make each other accountable, as the tasks can be assigned to specific students with a due date that they agree upon. As the software environment is so flexible, you can teach students to use Trello® as a brainstorming space that initially records all of their ideas. They can then break these down into steps later. As a teacher, you can then check in on how each group of students is going by simply visiting their board and seeing which tasks have been completed and what is left to do. You can put stickers on each task as rewards, too!

- **Assigning regular student tasks**

  There are always things that need to be done on a daily and weekly basis in a classroom for it to function properly. Why not record these tasks in Trello® and assign responsible students to them? Even simple things like watering the plants or feeding the hermit crabs could be recorded. This could be extended to include tasks from assigning the week's homework through to lists of what students could be doing in their school. Some schools might find this useful for planning tasks among teachers as well, including scheduling professional development activities, organising the school dance or even assigning the staff room job of stacking the dishwasher!

- **Recording links to resources**

  This is fairly obvious, but it still needs to be noted. You could create lists of science websites that you love, and attach in-house worksheets you've created. Maybe linking to Pinterest® boards you're curating would be a handy way of streamlining this.

- **Communication with parents**

  This one is powerful for improving communication between families and your school. You could share the semester's learning tasks, upcoming events or simply each week's homework.

As you can see, Trello® is an incredibly versatile and effective tool when it comes to education. No doubt you can think of other ways you could be using this app in your classroom too! As always, the outcomes you get will be highly dependent on how you introduce the program to your class and the subsequent support you then provide. Still, if you're looking for a classroom collaboration tool that is quick and easy to integrate, it's pretty hard to ignore this solution. Try it out; you might be surprised how quickly you get hooked!

The next section may seem like just a little bit of fun but it has a hidden agenda ... creating engagement with learners who don't normally participate. You know those memes you see posted and shared around the internet? Well, in your hands, they can become powerful teaching tools!

Teaching science through memes

'I like the app where you can make your own memes. I make memes all the time and send them to my friends.'

— Taylor Swift

- Can you teach in a language that students love and respect?

- If they're smiling ... they're engaged! Can you make them smile?

Who doesn't love a chuckle? When you enter most science faculty staff rooms, you'll undoubtedly notice a variety of humorous education and science posters on the walls.

Ranging from the mildly amusing to the downright hilarious, the internet is brimming with viral images that look at the lighter side of science – the side that people love to check out and share. Funny images with captions connect with people in a way that other media doesn't ... so why not use these as an engagement tool in your primary science classroom?

A meme is a viral image relating to a simple idea that spreads rapidly throughout the internet due to social sharing. Usually funny, the central quality of a meme is that people identify quickly with the meme's background image and the phrasing. As the viral image becomes instantly recognisable, people begin to make their own captions across the image to make its meaning apply to their own ideas. Everyone loves it and the internet is filled with memes as a result. This means that, as a teacher, you can use them as a tool to quickly convey a message to your class in a format that students will recognise and appreciate.

Take Grumpy Cat and Chemistry Cat ...

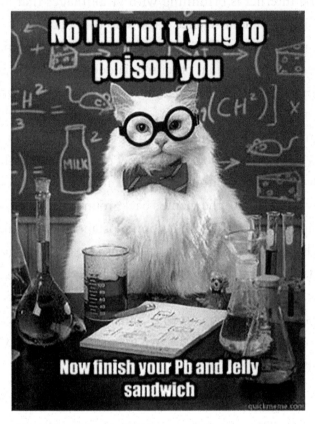

Chemistry cat meme via www.quickmeme.com

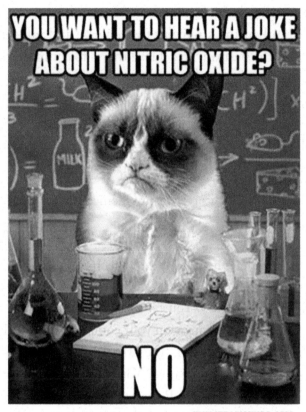

more awesome pictures at THEMETAPICTURE.COM

Grumpy cat visits the chemistry classroom via
www.themetapicture.com

In the above, you can see that the hugely popular Grumpy
Cat has been spliced into the Chemistry Cat meme. In both
cases, the meaning is grasped straight away and it provides
a talking point for students around the labelling of chemical
elements. Of course, you could have standard posters on your

wall, but where's the fun in that? As a teacher, you want to be constantly using strategies that engage and inform students, and the best way to achieve this is to communicate in a way that resonates with them.

So, how do you create a science meme yourself? There are a variety of meme-creating websites that will let you overlay a small sentence over a thought-provoking picture. Look up the following websites to get you started:

- Meme Generator®: https://memegenerator.net/

- Meme Creator®: https://imgflip.com/memegenerator

- Imagechef®: http://www.imagechef.com/

It's really quite easy! You can create a meme, create a free account and save the image on your computer in roughly one minute using Image Chef. The other sites seem fairly easy to use too. Not feeling creative? You can find science memes easily if you trawl through Pinterest® or simply use Google® image search. The only downside of this is that you might find yourself spending a lot of time searching for the appropriate image and caption combination; in fact, it can almost feel like you're Alice falling down the rabbit hole. Because of this, I recommend that you spend the time making your own memes and saving them to your computer for later use.

If you do use someone else's meme or cartoon, make sure that you give credit to the people who created it. Someone has spent time putting together the meme, so it's only fair that you reference what they did. If you're placing the image on your classroom blog, you should link to the original page where you found it and make a note of the website and creator. In fact, you should do this with any content that you find on the internet; in most jurisdictions, failure to do this actually amounts to copyright infringement, and the last thing you need is an issue like that when you're just trying to teach students! Let's be honest, I know that many people often skip this step, but we're aiming for best practice here. Besides which, it's simply a nice thing to do. As a side note, if you're unsure whether you can use an image, why not contact the creator and ask? More often than not, the creator will be more than happy for you to use the image and will simply ask for a link back to their site ... which isn't a big price to pay.

Once you get into the swing of it, you'll find that creating science memes becomes just another part of your skillset as a teacher. You might find that projecting a science meme onto your interactive whiteboard and having it work in combination with some experiment stimulus materials at the front of the room is a powerful way to gain students' attention. You could also hold a student competition to create the best caption for a meme, and showcase it with a science project or simply use it to summarise a learning topic succinctly. It's up to your imagination and comes down to the time you've got in your classroom to make it happen. Just be sure to use images that convey meaning without being insulting

217

– unfortunately, there are images out there that are derogatory to a person in the image. This is a sad reality of what can go viral on the internet. However, once you set up some ground rules, you'll find that this can be a bit of fun where the kids learn a thing or two in the process!

Action points

- Encourage students to start blogging about their learning. With the right platform and child protection measures, it can become a powerful reflection tool, as well as a vehicle to spread knowledge.

- Take advantage of the huge amount of information being shared on Pinterest®. Even if you're not keen on curating content for others, you can create private boards containing interesting experiments and science activities.

- Create a short-form science video fair and get students to submit fifteen-second videos to you for possible inclusion on your classroom site!

- Seriously consider Trello® as an organisational app for project-based learning. It is incredibly powerful for those willing to shake up how knowledge and thoughts are arranged and shared in school.

- Have fun adding science theories, quotes and ideas to viral memes. The students will love you for it.

Notes

6. Engaging the community

'Education is for improving the lives of others and for leaving your community and world better than you found it.'

— Marian Wright Edelman

S HARING KNOWLEDGE IS an awesome thing for students to do, and can cement their own learning in the process. If given the right structure, they'll really take it on and shine. With that in mind, let's see how you can take teaching to a whole other level by creating events that your entire community (local and global) can get involved in!

Create a school science fair that draws attention!

'Engaged parents can strengthen communities, mentor and tutor students, and demonstrate through their actions how much they value their children's education.'

— Arne Duncan

- How can you showcase just how well your students have been doing?

- Can you grasp the opportunity for both student and personal growth?

Science fairs are so much fun to put together and they really bring your school hall to life when done properly. Who wouldn't want to come visit your school when the students have created robots, DNA extractions, weather simulations, colourful chemical concoctions and more? It's certainly much more interesting than bare, polished floor space! And what better time to do this than during National Science Week, on World Science Day or in conjunction with another similar occasion?

The classic volcano science fair setup – St Patricks
Catholic Primary School, Kogarah

Science fairs naturally bring science to the community.
They also create a focal event for your school that can help
highlight all the things that are great about education. There
are so many things you could do:

- **Showcase your student's skills** in a pop-up Maker
  lab! As we've discussed, students love to deconstruct
  things and re-use objects in creative ways. With a
  little supervision, you can have students producing
  all manner of strange and wonderful objects to show
  off. Have students set up tables around the hall on all
  types of science, and have the community visit during
  the school day or in a special evening session. BBQ
  time perhaps!

223

- **Have scientists visit your school.** You'd be surprised what you can get if you just ask. No doubt there are engineers, surveyors, scientists, doctors, National Park Rangers, council environment officers and more from your students' families who would be willing to give up a little bit of their time to speak with your students and the community. You could also get in touch with the Tall Poppy Campaign, which works with a couple of hundred scientists from across Australia, all of whom have great communication skills and want to work with the community.

- **Hit the streets!** I remember bring involved in a fantastic science day that Gresford Public School held a few years ago that engaged local council and business owners to help hold their science fair in the community itself. It was one of those win/win/win/win situations … the school got to showcase their students' work, the science fair drove local tourism upwards for the council, the cafes and shops generated extra income from the increased foot traffic along with subsequent goodwill for helping host the event, and it exposed the greater community to science through the day's events and media articles. Gresford Public School worked in partnership with Dungog High School and invited local scientists, headed by local favourite Dr Leslie Wright, to help present on the street. The atmosphere in Dungog was tremendous. Plus, they got it funded by the Federal Government! Is this something you could make happen in your town or suburb? Why not give it a go?

- **Let your hair down!** Okay, in some ways, this is not politically correct, but students really do enjoy dressing up as mad scientists. Yes, I know that it perpetuates the long-standing concern that this image is unrealistic, as we discussed at the beginning of the book, but, at the same time ... it's meant to be a fun day! With enrolments in HSC and university science courses dropping, it is our duty as science educators to excite and inspire students in whatever way possible. If it means there will be energy and buzz in your school, get the kids dressed up as scientists. They might engage more with your lessons if they see that you can relate to them and not always be so serious. This is another opportunity to survey the students about what they think scientists actually do. Talk with your students about whether the crazy scientist caricature really exists and how this image has become entrenched in modern culture. You might be surprised how quickly you can turn their perceptions around by getting a little silly.

Teachers at Al Faisal College in Auburn letting their hair down!

As you can see, there is a lot you can do to make a fair at your school vibrant and engaging. Why not talk to your oldest primary students and get them to steer some parts of the project themselves? With your guidance, they will learn project management and community engagement skills, while learning science to boot. You could even build this into your ongoing assessment. What's not to like? Give it a go and school engagement will certainly benefit as a result.

Worldwide and national events also produce another source of inspiration for teaching science in primary schools. With a little planning, you can arrange a community event at your school that helps support a greater cause.

Create a community science garden

'Education is very important, and the botanical garden is the place to do that. I grew up in a semi-rural area and learned from that being my playground.'

— Nell Newman

- Is there somewhere in your school that could benefit from a new green space?

- Whom can you partner with to make this happen?

Students absolutely love getting their hands dirty in the garden. The traditional school garden offers a great chance to teach students not only about growing and caring for plants but also the importance of biodiversity, how agricultural science impacts their everyday lives and even local culture and customs. The best bit is that a garden is something everyone can appreciate, so you could put out a call to action in your school newsletter that you'd like to work with the community to establish a new garden in your school! I've seen schools put on working bees where, in one day, all of the plant beds get made, filled, mulched and planted with the help of families keen to earn their afternoon BBQ!

If you ask around your school, you might find arborists that are more than happy to donate excess mulch they have

227

from tree lopping. You may even find a timber store able to donate wooden beams and off-cuts for the garden beds, and a landscaping supplier prepared to sponsor the topsoil and fertiliser. In many cases, local businesses are more than happy to help and you could show your appreciation by showcasing them in your school newsletter. As always, if you make it a win/win situation, you're bound to have people ready to help you get students into the garden to learn about agriculture and the environment.

The following list offers some ideas that you could use to put some extra STEM learning opportunities in your school garden – just have your students wear gloves and hats for the job!

- Putting in a school vegetable patch offers a fantastic opportunity for students to create their own plots to experiment with. There are so many variables that the students could control: The plant species chosen, the light, watering schedule and more. The actual soil you put the plants in offers several variables you could control as well, such as the soil pH, texture and salinity. Your students could run a longitudinal experiment on how changing one of these variables above impacts upon plant height, leaf number, leaf colour, number of fruits grown or more. Additionally, your students could document their learning as they go via a classroom blog, and the whole class could be managed via an app such as Trello®. Don't forget to set up your experimental plots with clear labels, not only with the students' names but also the plant species, the plot number, the date of the trial and the variable being tested!

- Having a school garden offers the opportunity for students to learn plant propagation techniques used in modern plant nurseries. Why not see if you can work with your community to create a small glass-house? A quick trip to the local hardware store could give you a series of shelves, plant trays, plant root striking hormones and simple tools to get the job done. If your school budget is tight, you could also set up a mini-greenhouse using milk crates, cable ties and plastic sheeting. While you're at it, this could be a great opportunity to ask a horticul-turalist to share their experience with your students and give them deeper insight into how it's all done in the real world.

School garden at Warrawee Public School, NSW

- Simply walking students through the school garden can help you pick up a variety of student misconceptions on plants. As mentioned previously, it can be quite startling to learn that students often have mixed ideas in terms of what they think a fruit, drupe, seed, root, tuber or leaf actually is. Taking opinion polls in the garden will add an extra layer of fun to this exercise!

- You could use the school garden as a way of teaching the value of composting vegetable waste. Not only is this a useful way of teaching recycling, you're also getting the students to emulate processes found in nature. You could set up a worm farm too! Just be careful to have students wear facial masks when working with compost and to wash their hands afterwards.

- You could get students to set up rain gauges and anemometers to measure local climate conditions that your school garden is exposed too. You could use data loggers to make note of the ambient temperature and humidity levels and record the plants' responses to these abiotic factors.

- Apart from setting up your school garden patch to create an area for the classic plants found in gardens worldwide, why not put in some indigenous plants endemic to your area? Speak with your local council or bush care group to find out what plants you might be able to put into your school to improve the habitat for local fauna. You might even find that you could get free plants through National Tree Day. Try not to simply plant the largest, prettiest flowers, but instead create a mix of plants of different shapes and sizes that are known to associate together in your area's plant community. We've run several video conferences with schools to help classes set up their own ecological communities and you'll find that allowing students to create the school garden themselves will give them ownership of the site and, therefore, respect for it.

- Perhaps you could get students to keep a journal on the animals they find in the school garden. You could get students to look underneath leaf litter to find invertebrates scurrying around as well as watch where bees are travelling to and from (perhaps an opportunity to discuss how bees navigate using the sun!). You could also get students to take recordings of bird calls occurring in the area.

- Putting in endemic plant species into your school garden could raise the possibility of planting in native 'bush tucker' foods too. While you could source these plants from the local nursery, it might be even better to get a local indigenous elder to visit your school and speak to the students about traditional ways of using plants in the region – it doesn't have to be just about eating the plants; it can be about soap making, dyes, rope and more. It is important, how-ever, for safety that you stress that students should not eat plants found in the wild without an expert guide. A great example of partnership being set up be-tween schools and aboriginal elders can be found via the National Indigenous Science Education Program out of Macquarie University, which works closely with schools, indigenous elders and scientists to get stu-dents into science, as well as help scientists learn more about traditional methods of using Australian plants that could be employed in agriculture and medicines.

- The establishment of a school garden invariably means that you'll have to maintain it. Not to worry, though – this is another learning activity! Students are often not aware of the types of weeds that can be found in a gar-den and they can, therefore, learn different control tech-niques. This could also lead to discussion about the im-pact of weeds in local bushland and how they are spread (wind, water, animals or dumping). Great knowledge to have when taking your students out on a bush walk!

Even if your school doesn't have the room to put in a full-blown garden, you could still create a series of planting beds along the windowsill of your classroom (making for great classroom stimulus material!). Your students could also treat the garden as a mini-tourist trail by creating interpretative signage for the plants that are on display, as well as pointing out features such sedimentary rocks, logs with growing fungus, tree trunk hollows and more.

Observance days and STEM learning opportunities

'Saving our planet, lifting people out of poverty, advancing economic growth ... these are one and the same fight. We must connect the dots between climate change, water scarcity, energy shortages, global health, food security and women's empowerment. Solutions to one problem must be solutions for all.'

– Ban Ki-moon

- Students are deep thinkers who care about their world; how can they participate in it?

- How many STEM-teaching opportunities can you spot?

Want to create a buzz in your school? Perhaps the United Nations focus campaigns are a way to do it! There are so many global challenges that scientists, engineers, doctors, mathematicians and more are coming to grips with, and it's these very challenges that can really lift the attention level and, therefore, engagement of your students when described properly. Why not use these focus topics at your school?

I was spending some time scheduling video conference science lessons one day and noticed just how many

233

'International Days' can be found listed by the United Nations. Seriously, there are heaps of them! Everything from World Meteorological Day on March 23 to International Literacy Day on September 8, with links to a lot more information than you'd think possible. You can find everything to do with an occasion, from how to get involved, extra support materials, background history and fact sheets on why the topic is important to technical resources, including country profiles and campaign partners. A great example of this is World Malaria Day on April 23, where the campaign is clearly designed to inform the public about the current situation and what the World Health Organisation is doing about it.

So, what on Earth to do with all of these international campaign days? Well, like any science teaching and learning sequence, it would be logical to work out if any of these days fall within the time you're planning on running a unit and go from there ... on the other hand, why not flip this on its head and plan your lessons to take place when the days occur? As usual, this takes some planning to execute properly, but even if you only come close to the day itself, you can still leverage the campaigns to create STEM learning opportunities for students. Who knows? You just might teach the students that the science topics you're talking about have worldwide and real-time implications!

With this in mind, here are a few teaching ideas that you can easily implement if you decide that you want to use these international campaigns to create a sense of purpose and immediacy in your science lessons:

- Have students design posters, create a blog post, record a podcast or undertake a research project on a United Nations campaign topic. Not only does this create a clear student outcome that can be measured as an assessment unit, it will help the campaign itself gain traction with interested people, as your students will effectively be promoting the campaign cause. Of course, anything public needs to have school approval for the usual child protection and image copyright issues, and the content generated has to be scientifically accurate, but you can use this as a student learning opportunity too! How? You can clearly outline the requirements for scientific accuracy, fairness and attribution of intellectual property to your students – which are always required in scientific discourse, journalism and public information campaigns. Nothing like bringing the real world into the classroom!

- If the students are clearly excited about a particular topic, why not go all out and arrange an information night around that topic? If the students are prepared to put in the time and the resource requirements aren't too high, it could mean that your students will be more engaged with your unit of work. Plus, you just might raise public awareness enough to attract donations to the cause, or at least create some community discussion around world events at your school. A seed for creating a learning hub perhaps?

• Why not connect with a school from overseas and discuss the topic with students from another culture? As we've discussed, with video and web conferencing so accessible these days, it really is just a matter of turning on a device that's connected to the internet and getting the software to work. Time zones are not necessarily an issue either, as you could create a special breakfast session to meet another school that is having an evening BBQ – why constrain learning to between 9:00am and 3:00pm? The one important point is to use a time zone converter and run a few test calls with the remote site to make sure everything will work okay on the day. It might be worth connecting with other teachers using meet-up sites via the Centre for Interactive Learning & Collaboration (CILC®) or Collaborations around the Planet (CAPspace®) to see what they're up to.

• Another way to engage with a campaign is to ask a guest speaker to come to your school to talk to your students. As a subject matter expert, they could bring a wealth of knowledge to your classroom and allow students to converse with someone who is directly involved with the focus topic. This subject matter expert could be a doctor, a scientist, an engineer, a community worker, a science communicator, an author, a government representative ... the list goes on. Again, video conferencing could help here for guests that can't visit your school in person.

These are just a couple of quick ways to engage your students with issues of global importance. If it's good enough for the United Nations to be interested in it, perhaps we should teach our students to care about it too.

Find out more about the United Nations international observance days here: http://www.un.org/en/sections/observances/international-days/

Getting creative creates community

'Art is not what you see, but what you make others see.'

— Edgar Degas

- How can turn your school into a vibrant space?

- Can you leverage the work being done in your school in the media?

Let's face it, when you think about the diverse backgrounds and interests your attending families have, it is unlikely that a one-track community engagement strategy is going to grab the attention of everyone. To really build a sense of ownership and pride in your school's STEM outcomes, you should consider multiple approaches that are delivered across the school year and not just 'during science week'. Some ideas presented below might seem potentially time-intensive, but if you spread one or more ideas out over a semester, these art projects need not be seen as onerous but rather a chance for your families to let their hair down (and learn some science in the process!).

- **Make a science mural**

  Yep, get out the old paints, brushes and drop sheets to literally the deck the halls with your students' favourite science scenes. A hallway gives you a natural narrative of progression along time and space ... why not use the length of one hall to create a continuous mural spanning the entirety of universe, starting from 13.7 billion years ago and progressing through the periods, epochs and ages to finally arrive at now? You could initially start with the students and parents carefully measuring the length of the hallway and marking in all the time periods you want to address. Both students, teachers and the community could have a discussion on what each scene should include, as well as how they would like to incorporate useful facts and interactive stations along the hall. If you can't donate that much real estate to a science mural, is there a smaller area you could use? I remember as a kid growing up in Townsville our class got to choose an area of the school to paint an image of the coral reef which was a hit with the community.

An issue that this could raise, though, is that a large-scale mural in quite a few cases is likely to take up precious wall hanging space that could be used throughout the year for hanging other pieces of student work. So, an alternative to painting the hallway is instead hanging student-made posters up on designated noticeboards or other hanging spaces. Each class could use their designated area to create their interpretation of the timeline you're producing for a couple of weeks and then could revert to using this area for their own classes needs. Plenty of schools have boards set up in their school hall or outside their library to display student work too. Just be sure to have your school administration and fellow teachers on board with this; no point in causing undue stress and financial cost against something that isn't wanted. No matter what, if you do get people to help you out, make sure that you acknowledge their efforts and that they share in any notoriety you might achieve!

If you do decide to get a wall covered in paint, perhaps this is a great time to call on an artist in your community (ask around, there could be several!) who can then work with your students so that together they create the outlines, colours, shading, foreground and the background imagery.

Not that interested in creating a timeline? Why not consider walls depicting the following:

- Famous scientists and philosophers (in your own country as well as overseas)
- The periodic table
- Constellations
- The water cycle
- Renewable energies
- Marine, aquatic and terrestrial environments
- Endemic animals and plants
- Famous science and explorer photos (Watson & Crick's DNA photo or Buzz Aldrin on the moon)
- Local inventions and those from abroad
- Optical illusions and mathematical shapes

- **Science cake cooking contest!**

Who doesn't love the look of a ferocious looking dinosaur cake or a volcano cake with bubbling lava coming out of the crater?! It might sound a bit odd at first, but there is a variety of science learning outcomes when it comes to making a science cake. Beyond simply creating the classic image of a science scene, having your kids work to create science cake with their family allows them to explore their creativity, and the act of baking and decorating gives you plenty of chances to slide some science in too! Besides which, making a science cake with family and friends also creates the opportunity for children to interact with their families away from electronic gadgets!

Some kids might want to support their cake with a supplementary science experiment that goes with the theme of their creation, which is perfectly fine. You will have to make sure that parents know that it is not absolutely critical that their creation 'be perfect', rather that it is simply an opportunity for kids to express their interest in the world while learning to follow a recipe procedure. Be aware that some parents can get a little too competitive with these things, so you should create a scaffold of suggestions and some guidelines – in fact, it doesn't even need to be a competition if you feel that this doesn't fit the needs of your students.

Having said this, here's a small list of just some of the things you could suggest for students to consider when constructing their cakes. You could lead up to the showing with a series of prompting experiments to grab their interest too!

a) So, some students want to make a rocket cake? Awesome! What a great time to talk about how rockets work and maybe make a film canister rocket or a tea bag rocket with them to keep up the excitement afterwards.

241

b) Volcano cake is on the cards for some students? Great! This is by far the most popular of the science cakes you're likely to see produced and it gives you the opportunity to make the classic volcano experiment out of vinegar, bicarbonate soda, detergent, water and sand in class.

c) They're into the classic 'mad scientist image'? Why not discuss in plain language some of the real ways that scientists work? Let them know that science is a methodical and logical way of testing variables to discover how the universe actually works. Of course, no other scientist is celebrated in quite the way Albert Einstein is ... so why not make a science cake homage to him?

d) The kids are into construction projects? Why not make a Lego cake? We've seen Lego party cakes with blocks of cake stacked together, a Lego® mini-figure cake, even a Lego® Technics cake. You could discuss with your kids how Lego® forces you to plan ahead and creatively solve problems; to see this in action, you could set them a building challenge to be finished by the time the cake has finished baking! Lately we've been seeing quite a few Minecraft cakes, courtesy of the spectacularly successful game series.

Lego® can join cakes together! – Photo: Amanda Foxon-Hill

- **Dinosaurs are the order of the day?** What a perfect opportunity to talk about local palaeontology as well as the classic dinosaurs with your students, and then maybe create some fossil replicas as well! We've seen a variety of pretend fossil bones, pick axes and paint-brushes, prehistoric scenes and more.

e) Danger and mayhem your students' thing? Before making a radiation sign or a bio-hazard sign, why not talk about some of the risks that have been taken in the name of science so that we can enjoy our modern life? Think of the work done by the early pioneers in deep sea diving. What about Dr Marie Curie's work on radiation, which sadly took her life with radiation poisoning? Or the extreme risk taken by early astronauts or vulcanologists? These days, things are much more controlled; however, it is recognised that exploring the boundaries of human knowledge can still present dangers.

f) Why not depict the instruments themselves? Bubbling beakers are easy to create with a bit of fairy floss (aka cotton candy) coming out of the opening. Before you set the students the task of creating a cake, you could cover the importance of carefully measuring out the ingredients and how to avoid parallax error.

Got to love a bubbling beaker

- **Biological science takes their fancy?** I've seen people create forest scenes, coral reefs and even the anatomy of a plant. One of the most striking examples of a simple cake that 'grabs the eye' is the eye itself, great for a gross science party!

Would you eat this eyeball cake?

- The act of making a cake forces kids to follow procedural thinking. This is a critical skill for students. Everything from assembling IKEA® furniture to putting together a bicycle requires this skill, so why not give your students every opportunity to practise this skill in the real world?

Why not suggest that students set up a variable test with their cake by making two or more small cakes (or even an army of muffins)? In other words, change one part of the procedure and bake two or more cakes instead of one. Not only will your kids experience fairly testing a cake recipe, they just might stumble on a better recipe using the scientific method!

It doesn't have to be that high end when it comes to STEM learning. Younger students are constantly amazed by how colour mixes together and the very act of creating coloured icing is a learning opportunity!

- **Create a STEAM gallery**

Your school hall could be a fantastic place to hold an exhibition! Every now and then, you see schools exhibit paintings and sculptures. Why not invite artists to contribute to your exhibition and depict science themes? There's quite a bit you could play with here:

- Light, colour and sound

- Environmental issues

- The rise of technology

- The human body

- Our restless Earth

- Rube Goldberg machines and more

You could approach local businesses to sponsor prizes and invite government representatives and personalities to attend the opening or the awards night. Your school could leverage the opportunity to showcase itself in the media, plus you could use tickets to raise funds for your school's science resources (especially if you have a science show on stage to add an extra layer of interest for the community). You could also use the exhibition as a way of unearthing talented people who could contribute artworks to your school too. Your exhibition doesn't have to be elaborate; its real purpose is to bring people into your school to share knowledge with your students and vice versa! Art can be a fantastic way of driving STEM engagement. One of my favourite projects, which is still running at the time of writing, is Neural Knitworks – an initiative where families knit together woollen neurons in an effort to focus attention on brain health. These neurons, also known as nerve cells, are then sent in to be knitted together into a woollen brain! (https://www. scienceweek.net.au/neural-knitworks/). There is another similar project where people work to make coral reefs out of crochet, meshing mathematics and science to create intriguing artworks (http://crochetcoralreef. org/). Both of these creations could look awesome in your school foyer!

Citizen science

'If I can get some student interested in science, if I can show members of the general public what's going on up there in the space program, then my job's been done.'

— Christa McAuliffe

- What support is available for students to join real scientific investigations?

- Do you really need special equipment to contribute to world knowledge?

The rise of citizen science projects has helped working scientists extend their reach through simply having more people on the ground contributing to data. This is your students' chance to contribute to real scientific research and in doing so get a sense of genuine accomplishment, as they're doing something 'real'. Joining in citizen science projects not only teaches your students scientific thinking, it also teaches them to care for their world and create lasting links with responsible citizens working to make a difference. The following list outlines some of the projects available to your students. For simplicity I have listed mostly Australian sites, however a quick Google® search will have you finding plenty of programs in your neighbourhood … if there are none, perhaps you could establish one yourself!

- **Bird monitoring surveys and conservation activities**

  Simply get in touch with your local bird watching groups to find out what you and your students' families can do to help out. Students will not only learn to take accurate recordings of bird species they see in their own area, but also discover how connected their community is to the rest of the country as they see how migratory bird populations change throughout the year. **Check out:** http://birdlife.org.au/get-involved/citizen-science

- **Explore the Seafloor**

  This is an interesting project whereby people take photos of the seafloor and submit them to marine biologists to catalogue the organisms present. This project might be of interest to your budding snorkellers and it's a great way for students and their families to contribute to biological research. Of course, anything to do with water comes with safety risks, so you should insist that they collect data with experienced adult swimmers in areas where there are no rips or other strong water currents. **Check out:** http://exploretheseafloor.net.au/the-science/

- **Streamwatch**

  A long-standing project on water quality monitoring, your students learn the impacts of urban development on aquatic ecosystems as they measure biotic and abiotic indicators of the health of their local creeks and rivers. Supported by the Australian Museum, this program is fantastic for teaching biodiversity and more. **Check out:** http://australianmuseum.net.au/streamwatch

- **Feather map**

  Run by the Australian Nuclear Science & Technology Organisation (ANSTO), your students can collect feathers found near and within wetlands around Australia and send them in to nuclear scientists who'll analyse them in an effort to track populations in order to protect waterbird habitats. Not only easy to engage in, this project also teaches students that nuclear scientists don't just work on reactors and weapons. **Check out:** http://www.ansto.gov.au/Resources/DiscoveryCentre/EducationResources/Citizenscience/index.html

- **Galaxy explorer**

  Using a computer or a tablet and an internet connection, your students can help astronomers classify galaxies from data collected from a range of telescopes.

Big data is a big problem and so your students' help will certainly alleviate some of the load! The information collected will help scientists learn more about how galaxies grow and evolve.

Check out: https://www.galaxyexplorer.net.au/

- **Wildlife spotter**

  Another fauna citizen science project! This time, your students can pore over the millions of photographs of rainforest, woodlands and more in search of bandicoots, bettongs, malleefowl and more.

  Check out: https://wildlifespotter.net.au/

- **REEFSearch**

  If your students are lucky enough to live near a reef system, they may have the chance to out to the Great Barrier Reef. If so, perhaps they could help researchers monitor animal behaviours and distributions. Sun and science – got to love that!

  Check out: http://www.reefcheckaustralia.org/reef-search.html

- **Weather Detective**

  Search through old ships' log books to pull out weather information that can be used for climate change research. Combining science, history and sleuthing skills, perhaps this project might grab your kids' imagination?

Check out: http://www.weatherdetective.net.au/about/

- **SETI@home**

  The Search for Extraterrestrial Intelligence (SETI) project has generated enormous amounts of data since its inception and the first search in 1959 that needs to be sifted through. SETI@home is one of the first citizen science projects established whereby you donate the latent computer processing power of your computer towards searching through radio signals collected by telescopes for a chance to find intelligent extra-terrestrial life! This project offers the opportunity for you to discuss with students the possibilities of life beyond Earth and what the requirements are for life as we know it.

  Check out: http://setiathome.berkeley.edu/

... and the list goes on! There are literally hundreds and hundreds of citizen science projects that your class can get involved in; it's just a matter of being aware of what is available and choosing a project that you feel meets the curriculum needs of your students. With a little help and some adult supervision, your budding young scientists can participate in real science and contribute to their community and beyond. Furthermore, involving students in citizen science projects brings about the opportunity for your students to report their findings on a regular basis throughout the year at school assemblies, on your school website and social media, or even in the media, giving students the credit for their efforts that they all crave!

More support networks to check out:

- **Australian Citizen Science Association:** http://csna. gaiaresources.com.au/wordpress/

- **Atlas of Living Australia:** https://biocollect-test.ala.org. au

- **Australian Museum Centre for Citizen Science:** http://australianmuseum.net.au/ australian-museum-centre-for-citizen-science

- **Discovery Circle:** http://www.discoverycircle.org.au/

- **SciStarter:** http://scistarter.com/finder

Partner, partner, partner

The best way to ensure that your school remains the heart of your community is to reach out to businesses, industries, government sites and non-profit organisations to encourage them to get involved with your students' growth. In Australia, a great place to start is Inspiring Australia (http://inspiringaustralia. net.au/). Other possibilities include:

- Can your school adopt a planting activity in conjunction with local council? National Tree Day is an obvious example, but perhaps the council could do with some help on another project outside of the days that get the attention in the press.

- Would a local café be interested in letting your students set up a pop-up science table with experiments that could grab visitors' attention? Your students would get practice presenting science to the public and the café could use this as an opportunity to drive more attention to themselves as a business that gives back ... and sell more coffees in the process!

- Can you work with your local energy company and government to create energy efficiencies in your school that the students could monitor? Alternatively, can you arrange a visit to an energy company to learn exactly how electricity is generated?

- Perhaps your school could join the local bush care group and learn how to safely control weeds in local woodland and thereby help restore local ecosystems? Bush care groups tend to meet on a monthly basis and your students could take before, during and after photos as well as help survey vegetation to measure their impact.

- Can you work with a local movie cinema to arrange a special screening of an IMAX movie or a science documentary? Prior to the movie showing, you could invite local scientists to speak about their work and students could present what they have been doing as well.

- Is there a manufacturer you can visit? While I was writing this book, my editor Sara mentioned an interesting community partnership whereby during her school's anniversary celebrations, all the students worked on designing their own time capsule and were able to contribute ideas for possible contents that could be buried under the school for future viewing. There was a com-petition where the finalists were invited to go fortnightly to an auto manufacturer (who picked the winner and crafted the capsule) throughout a school term to learn about the end-to-end production process of car manufacturing and the efforts undertaken by engineers to make it hap-pen. Apart from witnessing the machines, robots and tradespeople working together, the students learned the importance of economics, as the auto industry was once the bedrock of her city, plus, historically, the manufacturers played an important role during the world wars.

ACTION POINTS

- Set up a science fair in your school and get the community involved! Don't be afraid to bite off a little more than you can chew: Invite guest speakers from a variety of institutions and get the students to make a real go of it. If you get your local government on board, you could make this a major annual event that you can take to the streets; it just takes willpower and the nerve to take action.

- Create a new garden in your school! Involve your community as much as possible on a building weekend and document its progress as the plants establish. Try to include native plants wherever possible and work with a bush foods specialist from your local nursery, bush care group or council to include edible plants.

- Consider creating a wall mural or a painted hall timeline depicting scientific scenes.

- Launch a science cake contest!

- Can your school establish a STEAM gallery?

- Are there citizen science projects that your students and their families could get involved in?

- Browse through the United Nations' list of international days to find causes that your students are interested in. Once you've got a topic, pursue it with as much vigour as the science fair ... only this time, consider hooking up with another class from around the world to share perspectives.

- Look for ways to partner with local businesses, industry, government and non-profit associations to get students engaged in STEM in the community.

- Visit Inspiring Australia to make connections: http://inspiringaustralia.net.au/

Notes

You can do this!

'*Education is the most powerful weapon which you can use to change the world.*'

— Nelson Mandela

WHAT A RIDE! From creating awesome role-play scenarios through to establishing annual science fairs and more in your school, there are so many ways that you can be the spark that ignites a passion for understanding how the world works – not only in your own students but also across your community and beyond. Why settle for the mundane when you have every opportunity to make a real difference in people's lives through creatively challenging your students' mindsets and misconceptions?

Throughout the book, there have been many practical ideas presented that you can begin to adopt in your classroom straight away. The first step would be to think about the needs of each of your students; what are their attitudes to STEM and what misconceptions might they have about their world? From there, think about how your classroom is set up to engage learners in science and weave in lessons that create excitement and enthusiasm around science in general while addressing the scientific method. I'd definitely create that 'science go-to box' of science materials and encourage other teachers to do the same! Think about how you mesh technology such as apps, robotics and more into your weekly classes and consider the possibility of establishing a Maker space at your school to encourage creativity and entrepreneurial thinking. Look at how you can leverage social media to encourage active sharing of ideas for other schools to discover, plus create links with schools and expose your students to specialist knowledge and resources found in museums, zoos, aquariums and more through video conferencing and science incursion visits. Finally, take steps to involve the broader community with the work you're doing in school through science fair events, citizen science activities and more. Even share your school's love of STEM with the media to garner more support and, potentially, funding! Above all else, share what you've learned with other educators ... work closely with your fellow teachers to really build a leadership team that inspires others to mirror your school's approach and, in doing so, create lasting change.

The world needs teachers like you who can create a scientifically literate public, people who not only appreciate the world for all its beauty and wonder but also make rational, informed decisions based on evidence. By being the inspiration for students and the greater community, you will become a thought leader others will emulate. This will produce more engaged students in the here and now, and also ultimately lead to greater take-up of undergraduate science and engineering courses in the years to come.

Think back to your first feelings when you entered school as student teacher, and how you were prepared to try anything to get those students actively involved in your lessons. This is where we left our friend Sally in the introduction of this book, but now she's starting to think differently:

Sally's just got through her first year placement with her challenging Year 4 class and is about to start afresh with a group of Year 3 students. She's weathered the storm of late night lesson planning and coming up with daily survival plans for classroom behaviour management. As she enters the new school year, it is now she needs the most support. Sally knows that with a new class comes a new opportunity and so, working with her colleagues, she begins to implement the ideas presented in this book in close consultation with her school executive. Step by step, class by class, her enthusiasm grows as her new Year 3 group becomes the envy of the school over the year ... behaviour management has become nothing more than background noise as the students are now completely focussed on their learning tasks (in fact, she now has trouble kicking them out of class at the end of the day!).

Despite only being new to teaching, she is now finding experienced teachers asking her for advice in teaching science and the greater community is constantly thanking her for her efforts. By the end of the year, Sally is the 'go to' person for STEM in her school and is being offered the role of science coordinator for the next year. Her school executive recommends that she start attending professional learning seminars on STEM teaching and her confidence further grows as she connects with like-minded educators wanting to make a real difference in their respective schools. With her growing network, she begins attending connected school events through video conferencing and even gets funding to transform one of the school hallways into a mini-science museum made by the student's themselves! The establishment of the mini-museum grabs the attention of the media and now she finds herself being interviewed as an expert in primary science education.

Sally begins speaking at teaching events and is now writing a book on how to teach primary science herself. Over the years, her work influences thousands upon thousands of students and she becomes more and more proactive in spreading the message that STEM, done well, can really make the difference.

As educators, we all have the opportunity to do the same. It's time to reinvigorate your love of teaching and bring about sustained active learning in your classroom. Make it real, make it inclusive, make it awesome. In doing so, you will change your students' lives, and, in the process, positively change your own life too.

Your classroom can be a glowing example of how to engage students ... now is the time for action. Just don't forget to have some fun too!

Support resources

About Fizzics Education

Fizzics Education was formed in 2004 as a specialist science workshop and show provider that now delivers engaging science lessons to over 300,000 students across Australia and beyond each year. Apart from delivering

hands-on workshops, educational video conferences and stage show performances at hundreds of schools, a

major role for Fizzics these days is holding teacher professional development sessions for motivated teachers looking to enrich their classroom teaching and take STEM in their school to the next level. A receiver of multiple national business and education awards, Fizzics Education has staff in Sydney, Melbourne, Brisbane and Canberra at the time of writing, and is often asked to deliver science programs in museums, zoos and other non-traditional environments such as movie cinemas, shopping centres, libraries, juvenile justice centres, hospitals, park lands, sports stadiums, retirement homes, university theatres and any other facility able to accommodate large or small groups of people. It's a fun job, but someone has to do it!

Come and visit us at www.fizzicseducation.com.au, where you'll find:

- Over 100 free science experiments written with the classroom teacher and a tight budget in mind.

- A forum for practising teachers to meet and share lesson plans and ideas on how to teach science effectively.

- Articles on how to teach science for every year group.

- Inexpensive classroom resources that work.

- Information on available school science shows and workshop incursions.

- Opportunities to attend teacher professional learning events and workshops.

Twitter: @FizzicsEd

Facebook: www.facebook.com/fizzicseducation/

Instagram: @FizzicsEd

Pinterest®: au.pinterest.com/fizzicsed/

YouTube®: FizzicsEd

Science apps for a variety of uses

The nature of apps is that some of these may well become out of date or even unavailable by the time you read this; however, if you get on iTunes or Google Play® as soon as possible, you may be able to use these in your classroom. Hopefully some of these stick around; each have a part to play in making science awesome!

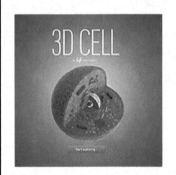

3D Cell by Invitrogen®
I love this science app! Yes, it's more advanced than primary students need, but it would be a shame not to know about this (you might need this for a gifted child, for example). Brilliantly rendered and designed with the educator in mind, this app will take your students into the living cell to highlight the variety of structures you can find, e.g. ribosomes, nucleus, cytoplasm and more. There are also images of the real thing and explanations on the structure and function of the cell organelles.

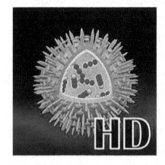

Microcosm

This science app is handy for kids to quickly understand the scale of things. Starting with lady beetles, progressing to viruses and ending with cosmic strings, the app is a great reference point for students. There is a small amount of information about each object as well. I'd use this in conjunction with taking students out to the oval with a tape measure to create the same scale.

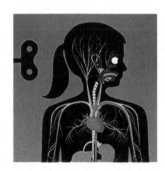

Human Body by Tinybop®

What a great science app! My eldest is in Year 1 and loves this app. The app is designed for you to peer into the human body and learn about how the heart works, how food moves through the body, how muscles work and more. While I suspect it was designed with the primary audience in mind, there is no reason that you couldn't also use it in high school.

Just Science by Novim®

Need a science app that depicts climate change? This app visually shows the temporal changes in temperature across the globe. Might make a useful animation for kids to investigate further. For example, you might ask: Why do baseline temperatures rise in Europe first?

Field Guide to NSW Fauna by Australian Museum

Museums all over the world are quickly pulling together science apps that allow naturalists and bush walkers to quickly identify plant and animal species while outdoors. Think of these as interactive field guides; you'll find animal distribution information, diet, habitat requirements, photos and sometimes audio as well. I recommend searching for field guides such as this for your local area and integrating them into your next environmental lesson outdoors.

WilderQuest by NSW National Parks & Wildlife Services

Explore rainforests, woodlands, heaths and more as you enter this virtual ecosystem. Created by the education team and park rangers at NPWS, your students have to find species within the landscape and learn more about their habitat and biology.

G2 Operation Aqua

This is an interesting way for students to explore looking after water catchments. Designed by SA Water®, the app looks at constructing a desalination plant to look after a town. If you build it correctly the population will be a happy ... if not, well, you get the idea!

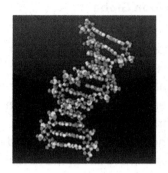

Molecules

As you might suspect from the name, this science app allows you to get a visual representation of a variety of molecules in a 3D environment. Bonding structures are all there, with the ability to zoom in as needed.

Pocket Universe

There are so many space science apps available these days! This one is handy as you can quickly use your device to work out the names of the stars and constellations above you using the virtual sky function. What's handy is that your location via the device GPS allows you to get a real time and quite accurate depiction of the sky above you. Highly useful for when you have your next telescope evening!

Moon Globe

Dedicated to the moon itself, this app allows you to explore the various craters, mares and more in a well-rendered 3D image. You can bring up the locations of the Apollo landing sites and learn more about the terrain features too.

Solar Walk

This science app allows you to explore the solar system in 3D. You can spend time visiting different moons and planets to get an idea of how the solar system is arranged and find out more information about the celestial objects you encounter. This is a highly popular science app for good reason!

Quakes

This science app gives you an up-to-date map of the locations of earthquakes from around the world. When you select each quake, you'll find out data such as the depth of the earthquake, its strength on the Richter scale, how far it was away from you (using the GPS information from your device), the time it occurred and where the source data is coming from.

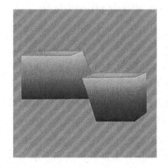

NSW Geology Map

Produced by the NSW Geological Survey, this map will allow you to explore the soils and rock types found throughout NSW. It would be interesting to use this in relation to the location of mines and townships as well as the vegetation types found in each area. If you're not in NSW, it would be worth looking on the internet for similar geological maps – you might just get lucky.

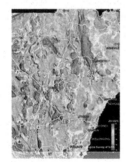

Earthviewer by BioInteractive®

This is a great science app that allows students to explore the changes in land masses over time due to plate tectonics. Not only can you see Pangea break up and the land masses spread apart, you also get to see changes in the oxygen and carbon dioxide levels, as well as the changes in day length due to slight variations in the Earth's orbit. Further information about each geological time period is given, as well as information on mass extinctions, the carbon cycle and more. A highly valuable teaching tool!

ExperienceScience®: Physics

This app has a number of animations that highlight exactly how energy changes and how forces interact in a variety of situations. Starting with the energies found with a skateboarder in a half pipe, you can then check out the physics behind a pendulum, the harmonics of a spring, how mass and weight work on a variety of planets, stopping distances of cars and more. A bit advanced for some students but worth having a look at.

Energy Island

This is an interesting science app where students can install a number of different electricity generators to power a town. Working to a budget, you can choose to install wind turbines, solar panels, geothermal plants and more. Once completed, you can run a simulation to see how efficient your created network is.

TinkerBox

This is an action and reaction type puzzler where students can place a variety of levers, pulleys, gears, springs and more into a situation to solve a problem. It's worth spending some time in the tutorial section to learn everything you can do, but you can certainly start messing around with the settings anyway. Worth a look as well!

VibSensor

This app allows you to measure, over time, the orientation of your device. Why? Well, by taking advantage of the accelerometer in your device, you can measure what happens in a collision or any other movement event. The app is effectively a data logger, creating a graph for when your device moves up, down, left, right, backwards or forwards. We had a bit of fun a while ago using a similar app while in a jumping castle! Time to get creative ...

Bridge Constructor

This is a fun little app where you can create all manner of bridges and see whether cars and trucks can cross over them. From a science point of view, it can test truss arrangements and whether a variety of suspension bridges can handle large loads. A bit of fun too!

281

	Poll Everywhere®
 **Poll Everywhere**	Why not set up polling in your classroom as part of everyday teaching practice? With Poll Everywhere®, you can get a very quick snapshot of a large crowd's thoughts on a given topic. A good way to use Poll Everywhere® is to create a poll and then put the website URL through a QR Code Generator. This means that if you put the QR code onto your presentation, students can quickly scan it and go straight to your poll. Otherwise, you could simply share the generated URL through your education platform, e.g. Ning.

Video conferencing resources

Centre for Interactive Learning & Collaboration: http://www.cilc.org

Collaborations around the Planet: http://projects.twice.cc/

ELB Education Events: https://www.education.electroboard.com.au/events/videoconferenceevents

Field Trip Zoom: http://www.fieldtripzoom.com/

Flat Connections: http://www.flatconnections.com/

Newsome, B (2014). *The Northern Districts Education Centre (Sydney) Churchill Fellowship to investigate best practice in science education via video conferencing.* Churchill Trust, Canberra.

http://www.churchilltrust.com.au/media/fellows/To_investigate_best_practice_in_science_education_B_Newsome_2013.pdf

NSW Distance & Rural Technologies: http://www.dartconnections.org.au/

Senior Learning Network: http://seniorlearningnetwork.com

Skype in the Classroom: https://education.skype.com/

Tate, J & Schultz, G. (2014). *Victorian Virtual Learning Project.* http://vicvirtuallearning.tumblr.com/

Virtual Excursions Australia: http://www.virtualexcursionsaustralia.com.au/

References

American Institute of Physics. *Operation Physics*. 1825 Connecticut Ave. NW, Suite 213 Washington, DC 20009 (202) 232-6688, http://www.aip.org.

Arizona State University. 2001. *Students Preconceptions and Misconceptions in Chemistry*, http://www.daisley.net/hellevator/misconceptions/misconceptions.pdf.

Bryson, B. (2003) *A short history of everything*. Transworld Publishers, London.

Brine Shrimp Direct. *Brine Shrimp in the Classroom*, http://www.brineshrimpdirect.com/c175/c76/brine-shrimp-classroom-c176.html.

Buxton, C. (2000). *Modelling science teaching on science practice? Painting a more accurate picture through an ethnographic lab study*. Journal of Research in Science Teaching, Wiley & Sons Inc, New York. 38: pp 387–407.

Commonwealth of Australia (2016). *National Innovation and Science Agenda website*, www.innovation.gov.au.

Hart, C. Mulhall, P. Berry, A. Loughran, J & Gunstan R. (2000). *What is the purpose of this experiment? Or can students learn something from doing experiments?* Journal of Research in Science Teaching, Wiley & Sons Inc, New York. 37: pp 655–675.

Drury, H. (1997). *How to write a laboratory report.* Learning Centre, University Sydney, https://sydney.edu.au/stuserv/documents/learning_centre/lab.pdf.

Fittel, D. (2008). *Reforming primary science education: Beyond the 'stand and deliver' mode of professional development. AARE 2008 International Education Conference 30 November – 4th December,* Brisbane, Queensland University of Technology.

Goodrum, D., & Rennie, L. (2007). *Australian school science education national action plan 2008 – 2012,* Canberra, Department of Education, Science and Training.

Knight, F. & Pizzey, G. (2005). *The Field Guide to the Birds of Australia,* Harper Collins.

Lehmann, K. (1996). *Bad Chemistry,* Dept of Chemistry, Princeton University, NJ.

Marbach-Ad, G. & Sokolove, P. (2000). *Can undergraduate biology students learn to ask higher level questions?* Journal of Research in Science Teaching, Wiley & Sons Inc, New York. 37: pp 854–870.

Oklahoma State. *Common Student Misconceptions,* http://www.okstate.edu/jgelder/acidPage25.html#Com

Patrick, H. Mantzicopoulos, P. & Samarapungavan, A. (2008). *Motivation for learning science in kindergarten: Is there a gender gap and does integrated inquiry and literacy instruction make a difference?* Journal of Research in Science Teaching, Wiley & Sons Inc, New York. 46: pp 166–191.

Phillips, N. (2014, October 6). *20-year decline in year 12 science and maths participation, study finds. Retrieved February 7, 2016, from The Sydney Morning Herald*: http://www.smh.com.au/technology/sci-tech/20year-decline-in-year-12-science-and-maths-participation-study-finds-20141006-10qvq2.html

Scherz, Z. & Oren, M. (2005). *How to change students' images of science and technology.* Science Education, Wiley & Sons Inc, New York. 90: pp 965–985.

Shepardson, D. Moje, E. & Kennard-McClelland, A. (1993). *The impact of a science demonstration on children's understandings of air pressure,* Journal of Research in Science Teaching, Wiley & Sons Inc, New York. 31: pp 243–258.

Roth, W. M. (1993). *Experimenting in a constructivist high school physics laboratory,* Journal of Research in Science Teaching, Wiley & Sons Inc, New York. 31: pp 197–223.

Williams, W (1980). *Australian Freshwater Life – Invertebrates of Inland Waters,* MacMillan Education.

Appendices

Colour blindness in the classroom

Many people are aware of the variety of colour blindness tests available that are based on the Ishihara colour blindness test plates developed by Dr Shinobu Ishihara in 1917. Often seen as the 'dot test', people can get an idea of how well they see colours by running themselves through a series of quick tests in which they need to pick out the numbers and patterns depicted. The thing is, how often does the inability to perceive colours get picked up in class? Does it really have an impact on education, or is it just an oddity? As I myself have a form of red-green colour blindness, I thought I might dive deeper into this from a personal perspective as well as from a science educator's point of view.

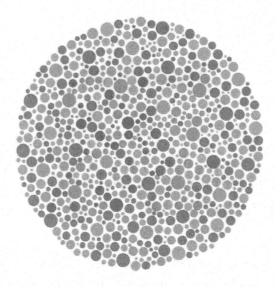

Just one of the many Ishihara Colour blindness tests
Image: Wikipedia

Only recently, I was at school running a human body science incursion and the topic of vision and colour blindness was clearly a major interest with the students. They were in the middle of a unit covering the five senses, and as I personally have red-green colour blindness, the kids were very interested trying to understand what I see when exposed to same frequencies of light as themselves. As the majority of the population sees the spectrum from red to violet, it is difficult for students to understand how colour perception can be different between people. Being a science teacher, I couldn't help but use my condition to my advantage, namely delving into the issue with real-life examples and how it actually works.

Do both sides of the photo seem to be the same colour? Perhaps this might indicate colour deficiency.
Image: Enchroma®

There are several types of colour blindness. My specific type of colour blindness is called deuteranomaly, which basically means that my retina at the back of my eye will not pick up much green light and instead is quite sensitive to reds, yellows and oranges. This means that greens, reds, yellows, oranges and browns all can all look quite similar to me, especially when it is dim. It also means that some blues and purples look similar as do some pinks and greys. Basically, the cone shaped structures within my retina have defective photopigments and, as such, I don't pick up light frequencies in the same way that the most of the population does. It's a sex-linked condition, in that the photopigment production genes are carried on the X chromosome and males only get one of those whereas females get both, causing males to show the condition more often (only one gene needs to be functional for normal colour vision to result).

You could imagine that once students find out that I'm colour blind, invariably they can't help but want to point out colours around the room in the hope to hear a strange answer ... yet most of the time, they are disappointed as I name most colours perfectly! You see, it is a standard misconception of students that all colour-blindness is alike and that we see things in 'black and white' (this condition, while it does exist, is extremely rare!). At this point, they're understandably confused, as they expect completely different answers. The reality is that five out of 100 males and one out of 100 females have my version of colour blindness, and we generally lead quite normal lives, in that it's not so much a complete lack of colour perception, rather just a difficulty in determining various hues.

But what does this mean for the practising primary teacher? Well, perhaps it's time to consider running one of these colour blind tests with your students. Let's be up front: We're not doctors and are not qualified to make a medical diagnosis. However, if you find that students clearly cannot read certain numbers in the Ishihara colour blindness tests, it would be high recommended that you at least let the student know and from there decide if the student is happy letting their parents know. Perhaps it might also change some of the ways you present visual information. Why? Well, you'd be surprised how it might be impacting your students. Here's just a couple of examples:

I can distinctly remember sitting in class during primary school, where often the only chalk left in the room was a dull pink colour. Using this on an old blackboard (which is really a dull green) made it very difficult to read anything. Unfortunately, at the time, I just had to squint and write down what I could, often to the detriment of my work. To this day, I still see noted speakers at conferences who have placed red writing on a green background in a PowerPoint presentation, without realising that they're making it really hard for some of their audience to read their words.

The issue can extend beyond primary school as well. In high school, we had to run chemical titrations using methyl red indicator. For normal colour vision students, this was not an issue, but for me, I couldn't see the distinct change between red to orange to yellow in different pH solutions, thereby completely missing the end point of the titration. This meant that when I had to enter in the volumes measured, I was always completely out. It wasn't really an issue in high school, as generally we worked in student pairs, however, in university, it was a problem, as the chemistry practical exam involved this same indicator ... bit of an issue!

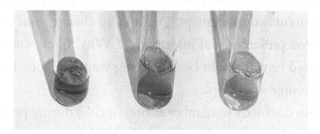

Methyl red changing colour in acidic to alkaline conditions (left to right)
Image: Wikipedia (user LHcheM)

These are just a couple of examples, but they're there to underline the point that students in your classroom just may not be perceiving what you think they are. Of course, it extends to future employment prospects, as in many jurisdictions, colour blind people are restricted from becoming pilots, electricians and undertaking certain defence jobs. Even if people with colour blindness aren't excluded, it can still hinder their work; imagine being a graphic designer working on a website, a painter or interior designer, a doctor diagnosing a degree of inflammation, or working as a pathologist trying to determine colour stains in histology! It's due to this that a careers advisor or high school needs to know who might have a colour blindness condition in your student cohort. If you're involved in the school executive, it might be worth considering some in-house training for your teachers on how to handle this, as although it may not seem that it doesn't impact day-to-day teaching, it actually could be, and your staff may not even be aware of it.

The recent development of Enchroma® glasses technology is interesting in that, in some cases, it may help to reduce the impact of colour blindness. There is some clinical evidence to support the claims and it could be very exciting for people in whom it has a positive effect! I personally have not tried the glasses out, but they sound interesting. Quickly, for full disclosure, I have nothing to do with the company nor any of their staff nor even any of their customers ... it just sounds like a cool technology that could be worth checking out.

So with all this in mind, it would be well worth you considering to trial some colour vision tests with your students. Even if none of your students turns out to have a condition, it would be a great talking point about the diversity of our genetic heritage and how sex-linked conditions work, plus it opens up a dialogue to discuss other conditions that impact people around the world (e.g. Hemophilia and Duchenne Muscular Dystrophy). Just be sure to be sensitive when it comes to how you go about introducing the topic, how you introduce the test and how you inform the student or family. After all, while, in many ways, most colour blindness doesn't impact greatly on kids' lives, it is still a condition that your student might be sensitive about – especially in the public arena that is your classroom.

How to improve school communication using Slack®

Anyone who thinks that a school is easy to run most likely hasn't worked in one! It's no wonder teachers have little time to eat their lunch considering the never-ending list of demands placed on them. Apart from keeping yourself organised (I recommend Trello® for that, by the way), you still have to work with your fellow teachers on shared project-based learning tasks, planning excursions, coordinating resources, meetings and much more. This can be quite hard when you only see your fellow staff members as you rush into the staff room for a quick bite to eat before heading out to lunch duty! As a result, many teachers have an overflowing inbox of emails from their fellow staff which they have to sort through themselves, as well as personal phones jam-packed with text-message conversations that can easily get buried.

Thankfully, there are many education apps that can make all the above much easier to deal with and one that just might change the game in your staff room is Slack®. This messaging app exploded onto the scene as a solution for busy teams needing a quick solution that could replace emailing and text messages with a unified area where quick messages between staff could be sorted into specific channels as they're being sent ... dramatically reducing the time needed to wade through messages for that all important urgent one. Imagine how useful this could be in your school; no longer do you have to sort and catalogue your messages into folders for future reference – this is already done for you by the messenger and, more importantly, everyone in your school can contribute to these team discussions using any device, effectively creating a highly accessible in-house forum that works. You can quickly check the channels that impact upon you the most, as well as send direct private messages to one or more of your fellow teachers. Slack® can really help your education team, with the option to integrate the online storage app Dropbox and project management tool Trello®, allowing your education staff to really work effectively together. The effect of implementation? No more inbox clutter, increased staff communication and better coordination of resources ... and the free version of Slack® does what most teachers would want anyway! Sounds like a win to me!

So, are you curious to try Slack® out for your staff? It can be quite easy to deploy on your smart phone, tablet and computer; it's just a matter of taking the time to set up the basics. First up, visit Slack® on the web, on iTunes or on Google Play®. Next, create a login to set up your team ...

You'll need to authenticate who you are using your email address. After that, set up the name you'll use ...

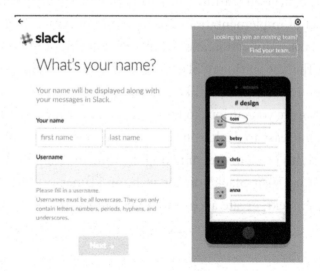

After setting your password and name, you'll need to name your team and come up with a web domain name ...

... and you're in! When you first enter the Slack® dashboard, you'll find an option to explore Slack® with the tutorials. What you will notice is the option to enable desktop notifications if you're using a computer – this can be handy to have on if you want to monitor information flowing among your team.

On the left-hand side of the app, you can see that you have already have been set up with two channels – 'general' and 'random'. You can now start adding multiple channels such as 'grade 1 science resources' or 'grade 5 mountain excursion' and invite people to join whichever channels you think they need to be involved in. You can describe each channel's purpose when you set it up as well.

It's so quick to get going! At this point, it's just a matter of using the app by sending messages back and forth between your education team (the more the better!). Very soon, you'll find yourself using email and text less and less, as Slack® becomes the go-to place for communication. As always, change management is an issue, so it would be worth you getting a few of your close teaching colleagues on board with the app first to trial it out and work out how you best want to use it. You'll be more likely to create enthusiasm in the rest of your staff room if people can see it working and how it'll help give time back in their already stretched schedules. Even though Slack® is a great solution for working together, no doubt your school already has a communication system in place and, as such, there could be people who will want to cling the old ways of doing things. Don't worry, though, in time they will see how much more effective you've become with your time and will want to join in too (especially when they see you can add silly random .gif videos in channels just for fun!).

Acknowledgements

Thank you to all my friends, family and colleagues at Fizzics Education and beyond who have helped me throughout the years; your encouragement and advice have been extraordinary. Seriously fun times have been had and I've immensely enjoyed this awesome ride with you all. A special shout out to Justine Bellamy and Holly Kershaw at Fizzics who've been the backbone of Fizzics Education for many years and to whom many in the company owe a debt of gratitude. Wherever and whenever possible, I'll have to thank you all with a bowl of liquid nitrogen ice cream!

I want to extend a warm thanks to all the science outreach community in the huge variety of museums, zoos, aquariums, galleries, parks, government representatives, companies and non-profit organisations that work so hard to get science out into the community. It's such a fantastic community to be a part of and your impact on science literacy cannot be underestimated. I'd also like to extend this out to my distance education friends throughout the world who continue to inspire students across the globe, with a special shout out to Jan Zanetis at CILC, Elaine Shuck at Polycom, David Foley out of NSW Distance &

Rural Technologies, Karen Player and all the crew at Virtual Excursions Australia, Amy Spath from CYNRIC and all the people in the Pinnacle Education Collaborative. Your work continues to inspire generations of people to want to learn more about their world... all of you rock!

I also want to thank Sara, Jacqui and all the team at Grammar Factory who have helped pull this book together for their efforts and professionalism. I know that the teachers and educators reading this will greatly appreciate your abilities in making this handbook useful rather than a tome collecting dust. Also, several people to which I'm very much in debt spent the time needed to make sure that this book fit the needs of classroom teachers; Jackie Slaviero, Christine Preston, Vanessa Barratt, Heather Catchpole, Jackie Randles, Jan Zanetis, Adam Selinger, Visnja Aw, Marc Wileman, Christine Draper, Paul Stafford, Sean Sullivan, SarahJane Dunford, Tim Reid, Nadeem Zreikat, Kirrilie Smout, Ben Minutoli, George Morton, Liz Fritts, Tony Peake ... much appreciated for your input, support, ideas and advice! Big thumbs up to Dave Faulkner, Aaron Tait, Summer Howarth, Louka Parry & Maddie Scott-Jones from Education Changemakers as well as Murray Deakin, Liz Fritts, Thomas Friedrich, Chelsea Cobb, Kirstyn Chan, Emma Shumack, Liz Cameron-Smith and Justine Felton at PwC Australia for their help in shaping Fizzics Education into the future.

Huge appreciation to the thousands of teachers and co-ordinators who bring Fizzics Education out to schools for school workshops and professional development events! Your continued support and enthusiasm for STEM continues to inspire us to create the kind of science outreach that students love to be involved in. Your constant commitment to guiding each generation to living informed and passionate lives is breathtaking.

Thanks to my kids Matilda and Jack, whose unbridled energy and inquisitiveness drive me ever forward to create more fun science activities that you will both love as you grow older. And, finally, to my ever patient and supportive wife Peta, without whom none of my efforts and madcap ideas could have actually come together. Since 2004 you've been there through every hurdle and every win, giving realistic advice and being a sounding board during the crazy times as this little company we started in the spare bedroom has grown and grown. Love you heaps!

About the author

Ben Newsome founded Fizzics Education in 2004 to deliver science education across Australia and beyond. Fizzics now reaches 300,000 students a year, with staff in Sydney, Melbourne, Brisbane and Canberra. Ben is a qualified science teacher, 2013 Churchill Fellow, 2015 and 2016 Australian Small Business Champion Education Services Winner and the receiver of several Pinnacle Awards from the Center for Interactive Learning & Collaboration.

Ben is on the leadership team for the International Society for Technology in Education Interactive Video Conferencing group, an Ambassador for the Association of Science Education Technicians NSW, and contributes to the education advisory committee for the GWS Giants AFL team as well as within the NSW Churchill Fellows Association committee plus the Center for Interactive Learning & Collaboration. In 2012, Ben co-founded Virtual Excursions Australia, which has now grown to include over forty major cultural institutions and departmental distance education managers, who create collaborative video conference events and share tips and tricks on best practice. In 2015, he co-founded the Pinnacle Education Collaborative, a network of over thirty cultural organisations in North America that deliver virtual excursions to schools.

Ben is a regular speaker at teaching and learning conferences and presents at teacher professional development sessions for K to 12 science class delivery. Putting together collaborative educational events with cultural organisations as well as the corporate sector is a major role these days, which he thoroughly enjoys!

He lives with his wife, Peta, and his two children, Matilda and Jack, in bustling north-western Sydney.

Twitter: @bennewsome_

Index

A

active learning 40, 264

B

Big History 106-108, 138
  Big History Project 107

C

citizen science 248
colour blindness 289-290
craft challenges
  Boat building 128, 130
  Building bridges 124
  Tower building 125
  Water bomb or egg drop challenge 122

D

discovery mindset 51, 54

E

entrepreneurial mindset 121
  pitchfest 109
experiments 13, 16-17, 19-20, 37-38, 40, 44, 46, 48, 55, 59, 69, 70, 77,
    82, 84-86, 91, 94, 101-102, 121, 133, 163, 171-172, 190-194, 197,
    200-201, 219, 241, 254, 268, 286

G

gifted children 42-43

M

Maker Movement 121, 132, 134, 136-137
  Maker space 131-132, 134-135, 137, 139, 145-146, 262
materials 9, 10-16, 38, 40, 43, 54-55, 69, 79, 82-86, 91, 93, 96, 111, 116,
      119-125, 127, 130,-132, 135-136, 139, 143, 164, 166, 171, 190,
      192, 193, 208, 217, 234, 262
  craft materials 16, 91, 119, 120, 124, 135, 139
  Lego 155, 242
  Meccano 56
  science go-to box 83, 85
  stimulus material 51, 54-56, 65, 91, 93, 232
misconceptions 15, 25, 26-27, 29-30, 31-34, 41, 47, 163, 230, 261, 285

N

natural world 113
  bush walks 113, 116
  plant structure 87

S

Scenarios 94
  space mission scenario 101
science communication 14
science fair 193, 222-223, 224, 257-258, 262
science garden 227
science models 15, 65-67
  solar system model 56, 176
scientist 20, 35, 36-37, 47, 76, 138, 225, 237, 242
  mad scientist 36, 37, 47, 242
social media 16, 189, 196, 198, 202-205, 253, 262
  blogging 190-192, 194-195, 219
  memes 17, 212-213, 216-217, 219
  Pinterest 206-207
  Slack 296-297, 299-300
  Trello 204-206, 208-211, 219, 228, 296

T

technology 4, 11, 16-17, 113, 119, 136, 141-142, 144-145, 173-174, 176, 177, 178, 189, 191, 193, 205, 246, 262, 287, 295
  apps 16-17, 117, 142, 144-145, 180-182, 184-186, 200, 202, 262, 270, 273, 275, 296
  coding 146, 151, 186
  drones 173-175, 177
  robotics 16, 102, 145-147, 150-151, 154-157, 186, 262
  video conferencing 16, 162-163, 165, 167, 186, 237, 262, 264, 283
the scientific method 4, 15, 38-39, 69, 70-71, 81, 91, 93, 133, 194, 246, 261
  conditions 38, 71-72, 118, 121, 129, 230, 294-295
  fair testing 70, 80, 194
  variables 38-39, 59, 70, 72, 75-77, 80-81, 91, 121, 194, 228, 242

V

Vine 194

Lightning Source UK Ltd.
Milton Keynes UK
UKOW06f1132010917
308379UK00006B/34/P